–THE–
BRITISH
HERRING
INDUSTRY

The Steam Drifter Years 1900–1960

CHRISTOPHER UNSWORTH

AMBERLEY

First published 2013

Amberley Publishing
The Hill, Stroud
Gloucestershire, GL5 4EP

www.amberley-books.com

British Library Cataloguing in Publication Data.
A catalogue record for this book is available from the British Library.

ISBN 978 1 4456 1081 8

Typeset in 10pt on 12pt Sabon.
Typesetting and Origination by Amberley Publishing.
Printed in the UK.

CONTENTS

Tables

Introduction:
The Herring Industry in 1900

Herring have been caught around the coasts of the United Kingdom for centuries, but until the early nineteenth century the size of the British industry and the methods it used had changed little from medieval times. The ports of East Anglia and the Firth of Forth had long enjoyed a thriving export trade in herrings but the volume they produced was small compared to that of the Dutch herring industry. By 1800 the Industrial Revolution was transforming many British industries and new technologies and production methods were bringing increased production, but the herring industry continued to operate much as it had for centuries. Change, however was on its way.

Nineteenth Century Expansion

During the nineteenth century, the hard work and entrepreneurial skills of the Scottish herring curers brought about a huge growth in the herring industry in Scotland. The curers packed freshly landed herring in salt into barrels – 'salted herring'. By keeping strictly to the quality standards set by the Fishery Board for Scotland and by using the 'Crown Brand' trade mark, the Scots developed a consistent product that was as good as the salted herring produced by the Dutch. Previously, the herring-hungry nations of northern Europe had bought very little British salted herring because of its poor quality; the Dutch product had always been preferred even though it cost three times as much. After 1860, the Scots curers began to take part in the East Anglian herring fishery and this gave a huge boost to the English herring industry.

Several technological developments contributed towards the growth of the herring fisheries, the first of which was the invention in the 1840s of machinery that could cheaply produce cotton fishing nets. These nets were larger in area but lighter in weight than the hemp nets which they replaced. By the 1880s it was estimated that by changing to cotton nets, some fishing boats were catching five times as many herring as they had with the old nets.

In the aftermath of a severe gale in 1848, which had caused the loss of many small open fishing boats, the Washington Report had recommended that fishing vessels should be of sturdier build and should provide fishermen with shelter from the elements. In larger, safer craft, fishermen were able to sail greater distances away from land, exploit new herring stocks and carry larger catches of fish back to harbour.

Some boat owners installed a small boiler on sailing drifters in order to power the capstan. This eased the heavy work of hauling in the line of nets and was another factor contributing to larger catch sizes – with a steam capstan you could work with a longer train of nets than could be hauled in by hand.

The Natural History of the Herring

Herring are an amazing species and understanding their life cycle is essential to catching them. They are a pelagic fish species, meaning that they live in the area between the sea bed and its surface, as opposed to demersal species such as sole, cod and haddock, which live permanently at or near the seabed. For much of the year herring live in large groups, 'shoals' or 'schools', in the deeper parts of the Continental Shelf. Once a year, shoals of adult herring (aged 3 years and above) migrate considerable distances to shallower waters to spawn (breed). Some herring groups spawn in the summer, some during autumn and some in the winter. Each of these groups has its own feeding and spawning areas and its own annual migration route. In the weeks before they spawn, herring feed up on the plankton that gathers just under the surface of the sea; it is the urge to feed that brings them into contact with the drift-nets which are set near the surface. In their pre-spawning state, herrings are at their highest nutritional value – the discerning Continental customers preferred their herring to be in this prime condition. After spawning, the 'spent' herring rapidly lose condition and are a less saleable product.

Herring shoals containing up to four billion fish have been recorded and a shoal of this size could be up to two miles in length. The shoals could generally be relied upon to re-appear in the same area each year, although occasionally they did not turn up or, if they did, the numbers were much smaller than usual. From very early times the three factors that made the herring an attractive quarry for fishermen were, firstly, their predictable arrival each year; secondly, their large numbers; and thirdly, they often came close to land.

Herring are an important part of the marine food chain. Natural predators include seals, dolphins, porpoises, whales, cod, hake, dogfish, seabirds and conger eels. Research in Norway has shown that an adult killer whale can eat 400 herring a day.

The Branches of the Herring Trade

Herring is a fish which deteriorates rapidly if it is not preserved by salting, smoking or freezing. By 1900, most of the British herring catch was being sold overseas as barrels of salted herring (also known as 'pickled', 'cured' or 'white' herring). This trade accounted for around eighty per cent of all the herring landed in the United Kingdom. Herring preserved by this method were still edible six months later. By the 1900s the British consumed very little salted herring, except in some Scottish fishing villages where it was a staple food in winter. Salted herring is an acquired taste – one less than enthusiastic British writer described it as like 'chewing a salty dishcloth, full of fish bones'.

Herring can also be preserved by smoking. Merchants in Yarmouth and Lowestoft had for centuries exported to Mediterranean countries hard-smoked herrings such as 'Red Herring'. These herrings might spend two to three weeks in a smokehouse, exposed to the smoke from slow-burning oak sawdust. British consumers preferred a lighter smoked herring such as the bloater or the kipper, which were smoked for just one day. A further product of this branch of the industry was the manufacture of savoury spreads such as bloater paste, once a favourite filling for sandwiches.

The arrival of the railways brought improved distribution networks and the demand for fresh herrings grew, particularly in the expanding industrial cities. In Lowestoft, the railway developer Samuel Morton Peto famously promised local fishing interests that if they supported his railway plans, the fish they landed in the morning would be on sale in Manchester the following day – and they were.

At the end of the 1890s, a new herring export trade was established with the German ports of Bremen and Altona (near Hamburg). This was the 'Klondyke' trade – so named (with a slight misspelling) after the Alaskan gold rush of 1897. It involved packing whole (ungutted) herring into large wooden crates together with a mixture of ice and salt. These were sent by fast steamer across the North Sea and when they were unloaded 24 hours later in Germany, they could still be sold as fresh herring or they could be further processed, cured or marinated. The Klondyke trade took large quantities of British herring (it accounted for 10 per cent of total herring exports) and, by providing competition to the Scots curers in the quayside markets, it helped ensure that the fishermen obtained reasonable prices for their herring.

From the early 1900s some manufacturers successfully produced herrings in tin cans, generally cooked in tomato sauce, and names such as 'Tyne Brand' of North Shields and 'Jenny' of Lowestoft would soon became household favourites.

A Geographic Advantage

At different times of the year, herring were found in commercial numbers in the North Sea, the Moray Firth, the Minches, the Firth of Clyde, the Irish Sea and the

English Channel. Because the herring grounds were close to land, British fishermen were generally able to land their catches of herrings within hours of catching them. Their French and rivals, who had to fish further away from their home ports, used larger drifters with a crew of up to thirty men (compared to the British drifter with a crew of eight to ten) which remained at sea for several weeks. As they hauled and emptied their nets, they commenced the salting and barrelling process on board the vessel. On arrival at their home port, the herrings were repacked and then marketed. The fact that the British herring were caught, gutted, properly salted and packed in barrels within 24 hours was the reason that Russian and German buyers preferred them.

Geography also accounted for the growth of Yarmouth and Lowestoft as herring ports. Being on the easternmost point of England, these settlements were closest to the migration and feeding routes of the great shoals of prime pre-spawning herring in the southern North Sea. Yarmouth's autumn landings were generally larger than those of its neighbour and in the early 1900s Yarmouth gained the title of 'the herring capital of the world'. Many of the companies based in these two towns had branches in other herring areas all around the country. The result of this is that, even though this book is telling the story of a national industry, Yarmouth in particular will feature strongly.

The Herring Year

At one time the British herring industry had comprised a series of separate, localised fisheries, such as Wick in the summer and East Anglia in the autumn. At the end of these two to three-month seasons, the local fishermen would switch to catching other fish species or just haul their boats out of the water. From 1860, as fishermen started to adopt larger, safer boats, they began to try their luck at more distant herring fisheries. If an owner had invested heavily in his boat and nets it made economic sense for him to get as much use from these assets as he could. This was the beginning of the traditional migration pattern of the British herring industry.

The year began in spring around the Outer Hebrides and then the industry moved to Wick, Orkney, Shetland, Fraserburgh and Peterhead for the summer months, fishing the Buchan stock of herrings. In late summer the industry focussed on the north-east of England (from Seahouses down to Bridlington), where the herring known as the Banks group congregated before spawning. Finally the industry moved on to East Anglia for the great autumn fishery exploiting the Downs stock of herring. This fishery was the largest and also the closest geographically to the Continental markets.

There were other, smaller stocks of herring around the British coast. After Christmas, some of the fleet headed west to Plymouth, Newlyn, Milford Haven and the Irish coast – all places where herring could be found at that time of year.

The large drifter-owning companies filled any gaps in the herring year by changing their nets and fishing for mackerel. Between each season, the fishermen returned to their home ports for a week or two for boat and net repairs.

The year-round nature of the industry and the fact that herring could be found on all British coasts meant that Lowestoft and Yarmouth boats could be seen in the Western Isles of Scotland, north-east England and the south-west; Scottish boats from as far north as Shetland could be seen in East Anglia, Yorkshire, Ulster and the Isle of Man; and Cornish boats could be seen in Yorkshire and the Isle of Man. In its own way, the industry brought together all corners of the United Kingdom.

Herring Ports and Herring Stations

At the end of the nineteenth century, the large English herring stations of Scarborough, Yarmouth and Lowestoft were prosperous ports whose economy was based on a mixture of tourism, sea-trade and both pelagic and demersal fishing. However, many of the British herring fishermen came from small fishing villages in the north of Scotland, in particular from the Moray Firth area, and for these communities their entire livelihood depended on herring fishing. The growth of the Scottish herring industry had lifted many people in these remote areas out

Preparing for the Lewis Fishing, Cullen Harbour

This turn-of-the-century picture shows the Moray Firth fishing community of Cullen seeing off the herring fleet as they prepare to sail for the spring fishing around the Isle of Lewis. It could be many weeks before they would be home again.

Downings Bay in County Donegal was one of several remote Irish herring stations where the British herring fleet and the Scots curers worked during the early part of the year.

of the deep poverty caused by the Highland Clearances, some of which were still happening as late as the mid-1800s.

In Scotland, herring were originally landed and processed at small herring stations established in remote settlements; because there were few harbours, boats often had to be hauled up on to the shore. As the fishing boats became bigger and their catches became larger, it made economic sense to concentrate herring landings on the larger ports. By 1900 it was safer and more profitable for fishermen to use the larger harbours because their drifters could enter them at most states of the tide and because they had good fish markets, better facilities and access to the railways. On the distant Scottish islands, however, seasonal herring stations still sprang up each year in remote locations such as Baltasound on Shetland, Castlebay on Barra and Stronsay on Orkney.

On the Scottish mainland, the harbours at small villages that had originally been established as fishing settlements were now deserted for much of the year because the local boats worked away from home. Their harbours briefly became hives of activity in between seasons as the boats returned for maintenance and net repairs before the fleet departed for the next season. In East Anglia too, boat owners in traditional fishing villages such as Winterton and Pakefield who had once landed fish on their home beaches now based their drifters in harbour at Yarmouth and Lowestoft.

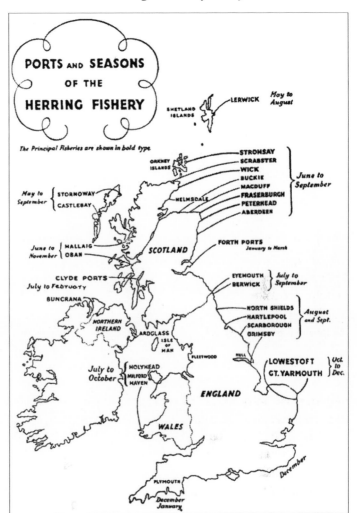

This 1930s diagram shows the different herring seasons and the main herring ports around the British coast.

Hopeman – A Scottish Herring Village

In Scotland some of the larger Moray Firth towns such as Fraserburgh, Banff, Buckie and Wick are well-known for once being herring ports but there were also a great many smaller towns and villages which also played a role in the industry. One of these is Hopeman, Morayshire, situated on the rocky southern shore of the Moray Firth, a village that was established during the 1805 Clearances. By 1855 there were forty-nine fishing boats in this village and 458 people were employed in the herring fishery as fishermen, coopers, curers, gutters and packers. In 1863, 10,000 barrels of herring were landed and packed at Hopeman, ranking it in sixth position in the whole of Scotland (although a long way behind Wick's 90,000 barrels). The village had its own salt store and an ice house. Such were the expectations of further expansion of the herring trade that in 1892 the Highland Railway built a branch

The old Ice House at Hopeman, Moray – a very impressive entrance to an underground chamber (as a small boy, the author used to imagine it was an Egyptian tomb!).

line linking the village to the main rail network and in both 1896 and 1897 over a thousand tons of fish were despatched from Hopeman by rail.

By 1900, Hopeman herring boats fished a long way from home and were landing their catches in Wick, Buckie and Fraserburgh; they travelled north in the summer to Shetland and south in the autumn to East Anglia. Hopeman was a one-industry village where very few of the men worked anywhere other than at the herring fishing and we will meet some of them as this book progresses.

The Economic Structure of the Herring Industry

In East Anglia the move to building larger vessels had resulted in a reduction in the number of family-owned boats. A new sailing drifter cost £800, which at today's retail values equates to £60,000. It required more capital than one fishing family could find from their own resources, so limited companies were formed to buy and run boats. The larger companies found that to make the most of the herring trade, it was not sufficient to just catch the fish. They needed offices at fishing ports during each season in order to support their boats, to sell their herrings and, increasingly, to manage their own in-house teams of curers and coopers. In Scotland, despite the formation of some limited companies such as the Caithness Steam Fishing Company (Wick) and the Steam Herring Fleet (Aberdeen), the majority of fishing boats remained family-owned.

At the end of the nineteenth century, anyone unconvinced about the growth of the British herring industry had only to look at recent statistics from the Ministry of Agriculture and Fisheries:

An industry that in just four years had increased production by 38 per cent and exports by 68 per cent was surely worth investing in. The value of exports in 1900 would equate today to well in excess of £200 million, giving an idea of how much the herring industry contributed to the wealth of Victorian and Edwardian Britain. This rapid expansion had, of course, been achieved at a time when the majority of the fishing fleet was still powered by sail.

There were good seasons and poor seasons. Even in a good season there were days when very few fish were landed and shore staff had no work. Sometimes the herring would be of poor quality, making it difficult to sell them in Europe. On other days

Year	TOTAL UK LANDINGS		TOTAL UK HERRING EXPORTS	
	Quantity	Value	Quantity	Value
	Hundredweight	£	Barrels	£
1897	4,993,500	1,282,000	1,119,200	1,364,000
1898	6,749,200	1,422,000	1,742,400	1,897,000
1899	5,761,500	2,006,000	1,743,000	1,899,000
1900	6,168,500	2,248,000	1,524,400	2,329,000

Table 1 UK herring landings and exports, 1897–1900.

there were huge landings which would cause quayside prices to collapse, forcing late-arriving skippers to throw away tons of perfectly good fish because no-one wanted to buy them. At these times farmers could for a few pence buy a cartload of good edible herring, which they spread on their fields as manure. So, at the end of the nineteenth century, despite the success and expansion of the herring industry, there remained an element of risk and a hint of speculation surrounding it.

As in any other industry there was tension and competition between the various sectors of the industry. Figure 2 shows the structure of the industry.

The five groups of herring buyers were all in competition to buy herrings as cheaply as they could. Fishermen were not bothered who bought their herring as long as they received a good price but they often felt that they were being diddled at the auction. Similarly, the buyers often thought that the fishermen were trying to slip some sub-standard herrings in among their catch.

Regulation of the Herring Industry

In England there was little government regulation of the trading and export of herrings – it was left to market forces, which worked well enough as long as there were plenty of buyers and plenty of herring. The government was concerned, however, about safety in the fishing industries – even as late as 1893, over 2,000 fishermen and boys had been lost at sea in one year. Various Acts of Parliament

An Overview of the Herring Trade

Boat-Owners
and
Fishermen

Catch and land herrings

Port
Authorities,
Salesmen,
Auctioneers

Sell herrings on behalf of the fishermen and take
their commission

Freshers	Smoke Houses	Scotch Curers	Klondykers	Canners

These groups all compete at auction to buy
herrings. They process them and sell them on,
taking their profit.

UK Fish Wholesalers	Herring Exporters	Continental Agents

Arrange transport and sell them on to the retail trade

Fish Retailers

Display the fish and sell them to the consumer

Table 2 Diagram of the herring industry.

Stencils such as this were used to mark barrels of Scottish salted herring. The Crown Brand was a guarantee to Russian and German merchants that they were purchasing a quality product.

had introduced compulsory registration of British fishing boats and competency examinations for their Skippers and Mates.

In Scotland, the Fishery Board employed district inspectors to monitor the herring industry by checking net mesh sizes and the structure and capacity of barrels. They were also responsible for the accurate recording of statistics and operating the Crown Brand system, the mark of quality which maintained the foreign buyers' confidence in Scottish salted herring. The Board's regulations and strict control had helped put the herring trade in Scotland ahead of the rest of the United Kingdom.

HERRING PEOPLE: Andrew Bremner

Andrew Bremner, herring curer and herring merchant, could look out with pride across the bustling harbour of Wick, in the far north of Scotland. Every summer, hundreds of fishing boats sailed in and out of Wick's busy harbour, landing herrings. Along with the local vessels, there were boats from other parts of the east coast of Scotland, from East Anglia and even from the Isle of Man, all taking part in the annual bonanza. By 1899 steam drifters – including the *Content*, owned by Andrew – were beginning to appear among the traditional sailing drifters.

The bustle and noise around the harbour spilled out into the curing yards where hundreds of young women, many from the Hebrides, toiled all day gutting herrings and packing them into barrels. Barrels of herring were stacked high around the harbour, waiting to be loaded on cargo ships to be carried to Baltic ports. Andrew owned a curing yard and a rope works next to Wick harbour and by 1900, at the age of 48, he had become one of the most respected figures in the British herring industry.

After Thomas Telford had built a harbour in 1806, Wick had grown to become the centre of the new Scottish herring industry. If pressed, Andrew would admit that the Wick season had now been overtaken in size by the great autumn herring fishery in East Anglia but he pointed out that Wick was where the modern herring industry had its origins. Andrew's attention to detail and business skills led to his name being recognised internationally for the excellent quality of his salted herrings and he now exported to Russia, Germany and the United States. Each autumn he travelled south to run a curing yard at Great Yarmouth. He was a man with no professed religion but was driven by a strong work ethic, serving as a Wick town councillor and magistrate.

HERRING PEOPLE: Richard Irvin

On the River Tyne, Richard Irvin's office overlooked the fishing port of North Shields. At 47, he was already a mogul of the fishing industry. He had begun in it at the age of 11 and worked up to owning his own company. He was involved in the early development of steam power in the trawling industry and by 1900, his company, Richard Irvin & Sons Ltd, owned thirty steam trawlers, ran a fish sales business and owned a Tyneside shipyard as well as running a fleet of whalers based in South Georgia. His company had offices in Aberdeen and at many other fishing ports.

CHAPTER 1

1900 to 1910 –
Continued Growth

The expanding British herring industry of 1900 was an exciting environment to work in. It employed a large, mobile workforce of men and women and it was part of an international trade that had links to much of northern Europe. Work in the industry was not just the manual labour of fishing, gutting and coopering; there were numerous salesmen, herring curers, merchants, exporters and clerical staff. At a time when few people in this country ever left their hometown, herring workers regularly travelled to other parts of the United Kingdom. However, this meant having to be away from home for months at a time, so young children would see very little of their fathers. It was said that because of these long absences, driftermen rarely knew anything about gardening or how to decorate a room.

The economic importance of the industry in 1900 is illustrated by the fact that the United Kingdom was landing more than 50 per cent of the total European herring catch. In Scotland, herring landings accounted for 50 per cent of total fish landings, showing how crucial the herring was to the Scottish fishing community (the corresponding figure for England was 14 per cent).

Steam Drifters Arrive on the Scene

Up to 1900, the growth of the industry had been achieved using sail power. However, there had been many instances where sailing drifters with a hold full of herring headed for harbour, only to become stranded for hours by lack of wind, too much wind, fog or unfavourable tides. Even on a cool day this delay could have a detrimental effect on the condition of the catch on board, and on the price received for it. Steam-propelled drifters were a means of avoiding such delays but there had been reluctance among herring fishermen to adopt steam-propelled vessels (one of the reasons they gave was that the noise of the engine would frighten away the herring). There had, since the 1870s, been various experiments in Scotland with fitting a steam engine into a sailing drifter and several steam drifters were built but not universally accepted by the industry. The turning point seems to have been the success in 1897 of the English prototype steam drifter the *Consolation* of

Lowestoft; this convinced the doubters and led to increasing numbers being built for East Anglian owners. Table 3 demonstrates their rapidly increasing popularity. Of the English drifters, some 90 per cent were registered in either Yarmouth or Lowestoft and the remainder were almost all from North Shields.

Year	Number
1899	16
1900	47
1901	105
1902	176
1903	226
1904	248
1905	262
1906	288
1907	346
1908	385

Table 3 Steam Drifter Numbers in England.

The Scots fishermen too were impressed by the steam drifter and Scottish boat owners soon began to place orders. By 1904 Andrew Bremner of Wick owned a fleet of five, and by 1913 the total number in operation in Scotland had reached 900 – a higher total than in England.

The demand for steam drifters led to a boom period for British shipbuilders. Many drifters were built over a period of thirty years in the larger ship yards of Yarmouth (200), Lowestoft (460), Tyneside (116) and Aberdeen (132). Some, however, were built in modest yards in small fishing towns and villages such as Berwick, Shoreham (Sussex), Porthleven (Cornwall), Peel (Isle of Man), Lossiemouth (Morayshire) and Hopeman (just the one, in 1909). The earlier versions were built with wooden hulls but within a few years, more expensive steel-hulled drifters were being constructed.

Steam drifters were on average 25 metres long, 6 metres wide and had a gross registered tonnage of between 80 and 90. The early ones were fitted with a compound steam engine capable of producing from 15 to 30 horsepower but later designs were fitted with a triple expansion engine which could produce up to 40 horsepower. Top speed for steam drifters was between 10 and 12 knots (1 knot = 1.15 miles per hour) but they rarely ran at top speed for any length of time because of the expense; they consumed less coal by travelling at steadier speeds. They had a secure, covered fish hold which could accommodate catches of up to 45 tons of loose herring.

A steam drifter needed more space on board than a sailing drifter. They needed a boiler room, a water tank and coal bunkers – they could burn up to ten tons of coal per week – so they had to be constructed larger than sailing drifters. Some drifters could carry over 30 tons of coal and 8 tons of water. They needed two more crewmen than a sailing drifter, an Engineer (or Driver) and a Stoker. The cost of paying these additional men meant that bigger catches were required in order to

Richard Irvin & Sons, Ltd.

NORTH SHIELDS,

With which is Amalgamated

W. H. LEASK,

FISH AND HERRING SALESMEN,

SHIP CHANDLERS, &c.

North Shields, Aberdeen,

Peterhead, Gt. Yarmouth.

Other Branches—

Blyth, Milford Haven, Lowestoft, Lerwick, Baltasound,

Stromness, Scrabster, Stornoway, Castlebay, Mallaig,

Lossiemouth, Downings Bay and Buncrana

Special Attention given to Drifters

FULLY QUALIFIED SUPERINTENDING ENGINEERS,

This 1913 advertisement for Richard Irvin Ltd shows both the variety of services offered by the larger fish merchants and the fact that they followed the herring industry around the British Isles.

remain profitable. The running costs of sailing drifters (luggers, Fifies, Zulus etc) were low, so it had been easier to make a profit from an average night's fishing. Sailing drifters, however, were still very practical and economic to run and some continued to be used well into the 1930s.

On average it cost around £3,000 to build a steam drifter (over £250,000 at today's values). Despite the large amount of capital required to buy one, the steam drifters were soon acknowledged to be a success and every boat owner felt they had to have one. As a result of the investment in steam drifters, herring catch sizes grew year on year in both Scotland and England.

HERRING PEOPLE: Richard Irvin

Richard Irvin, being a trawler owner, had not previously had much involvement in the herring industry but he watched with interest as steam drifters began to appear. In 1900, persuaded by the large catches they were landing, he set up The East Coast Herring Drifter Company and began acquiring his own fleet of steam drifters. In 1902, he would establish a fish canning factory on Tyneside producing tinned herring under the label 'Tyne Brand', thus ensuring that his fleet of herring drifters never had to dump an unsold catch.

Conflict Between Drifting and Trawling

The driftnet had long been recognised as the best net to use for catching herring without damaging or crushing the fish. Drift-netting is a passive form of fishing in which a curtain of nets is suspended just below the surface to 'drift' with the tide and currents and wait for shoals of pelagic fish species, such as herring, mackerel and pilchards, to rise to the surface to feed on plankton. The fish collide with the net and become caught by their gills in the mesh (similar to a boy getting his head stuck in railings). Trawling requires continuous forward propulsion, dragging a heavy net along the sea bed and catching whatever happens to be down there.

In the early 1900s, some of the trawling fleet began to fish for herrings and this move raised a number of concerns among drift-net fishermen. Firstly, any trawling in herring spawning areas dislodged and destroyed the new herring spawn and larvae as they lay on the seabed; the drift-netters feared that this could be reducing the future population of herrings. Secondly, being quite a small fish, herring were easily damaged by being crushed by the weight of the catch within the trawl net. Split or burst herrings did not appeal to European buyers; the driftermen worried that this inferior product might harm the good reputation of British herring, leading to lower prices. (At around this time, trawled herring were selling at Hull for 5 shillings a cran while driftnet-caught herring were fetching 42 shillings a cran). Lastly, the herring caught in trawl nets included many immature fish that were too small for the market and so were thrown back (dead) into the sea, again causing concerns over future stocks.

Feelings on the subject of herring trawling ran high. In 1902, when an Aberdeen trawler made the first ever landing of trawled herring at Wick, her skipper was stoned by local fishermen. The argument about trawled herring was to rumble on for some years.

The 'Last' Farewell, 1908

For centuries, the size of herring catches had been measured by number – the fish were manually counted according to the following system:

4 Herrings = 1 Warp
33 Warps = 1 Hundred
10 Hundreds = 1 Thousand (1,320 fish) = 1 Cran
10 Thousands = 1 Last

The Scots, who first moved herring fishing on to a truly industrial scale, had long since abandoned counting each individual herring. Instead they had adopted a simple volumetric measure, the cran (defined as 37.5 imperial gallons), as the standard quantity by which to sell herring catches. In 1908, Great Yarmouth, along with other English ports, adopted the cran measure for use on the Fish Wharf and a further step to modernising the English herring industry was achieved. Because of this move, Yarmouth, Lowestoft, Grimsby and North Shields all began to use quarter cran baskets for unloading herring catches. In East Anglia this generated a rush of work for basket makers as 7,000 of these baskets were initially ordered for Yarmouth and 3,600 for Lowestoft. The baskets all had to be regularly checked by Weights and Measures Officers and any that were inaccurate or damaged had to be replaced.

Although the cran measure was used widely within the herring ports, when it came to producing annual statistical returns on the fishing industry, both the Ministry of Agriculture and Fisheries for England and Wales and the Fishery Board for Scotland recorded all herring landings in hundredweights (a cran equated to 3.5 hundredweight). On an international level, statistical information on herring catches became more complicated; Norway used a volumetric method, the hectol; Denmark a numerical system, the score; and Holland the metric tonne.

What did the Scottish Curers Actually Do?

The Scots curers had revitalised the herring industry in the nineteenth century but how did they operate? Curing in itself was not a complex process but it entailed an element of risk and speculation. It required large numbers of skilled staff but did not need a large investment in buildings or heavy machinery. During the nineteenth century many Scots set themselves up in business as curers. In the early years they tended to be small-scale, part-time operators such as fish merchants, shop-keepers, coopers and publicans but, as the size and geography of the herring industry increased, curing became year-round work and curers established their own limited companies.

The raw materials a curer needed (apart from herrings) were barrels and salt. His labour force comprised coopers and women to gut and pack the herring. Originally, the coopers had constructed their own barrels but by the 1900s factory-made barrels were becoming available and the cooper's role became one of supervising the gutting (or 'gypping') and packing of the herring and then using his skills to seal the filled barrels.

The curer had first to hire premises, a curing yard or 'pickling plot'. The plots were usually just areas of open land close to the shore but in the larger ports some yards were more established and they had permanent buildings on them. In 1900 Yarmouth Borough Council raised £869 from curers by renting out sixty-one pickling plots on the open land of the South Denes, close to the Fish Quay. The curer ordered a stock of empty barrels and good quality salt – any impurities in the salt could discolour the herrings. The main equipment he needed was some 'farlans', trough-like work-tables where the gutting took place, and a means of transport – usually a horse and cart, but later a motor lorry. The yard was made ready with farlans and new barrels when the gutters arrived, a day or so before herring landings began. All they needed now was for the fishermen to start landing herring.

Curers bought fresh herrings each day at the quayside auction. Once delivered to their yard, they were tipped out on to the farlans for the gutters to start work on them. The curer continued to buy herrings throughout the day depending on their availability and price – the cheaper he could buy them, the more profit he could make when selling them.

Before the start of a season, the curers borrowed working capital from the Scottish banks to finance their purchases of herrings and barrels. Sometimes,

Gutting herrings, Gorleston-on-Sea. Gorleston quay was opened as a means of easing the overcrowding across the river at Yarmouth. Gorleston had its own herring salesmen, curing yards, coal merchants and modern transport facilities (!).

German buying agents would allow them an advance on sales. At the end of the season, when, hopefully, all their herring had been sold and the gutters and suppliers paid, the curers settled up their loans and credit. Poor quality salted herring were difficult to sell so it was important that the girls worked to the requirements of the Crown Brand product descriptions (Table 4). These categories were based on the requirements of the European buyers and show just how seriously the Continentals regarded their herring. The coopers were constantly on the lookout for any careless gutting or packing.

As long as there was a steady supply of herring and a strong demand from the buyers, the curers could make a lot of money. A poor herring season, however, could lead to spectacular bankruptcies, such as in 1888 when James McCombie, a curer in Peterhead, was found to have liabilities of £86,000, of which some £73,000 (£6 million at today's values) was owed to the banks.

The Crew of a Drifter

A steam drifter generally carried a crew of ten, led by the Skipper and his second in command, the Mate. Other roles included the Driver (or Engineer), and (on East Anglian boats) the Hawseman, the Whaleman, the Fireman (Stoker), two Younkers and the Cook. In Scotland any crewmen other than Mate, Engineer, Cook and Stoker were generally referred to as 'Deckhands'. The important role of cooking

Category	Definition
La. Full	Large herrings full of milt and roe and not less than 11¼ inches long
Full	Herring full of milt and roe and not less than 10¼ inches long
Filling	Maturing herrings of not less than 10¼ inches long with long gut removed
Mat. Full	Herring full of milt and roe and not less than 9¼ inches long
Medium	Maturing herrings of not less than 9¼ inches long with long gut removed
Mattie	Young maturing herring of not less than 9 inches long with the long gut removed
La. Spent	Spent herrings not less than 10 inches long

Table 4 Crown Brand categories.

the crew's meals was usually put in the hands of the youngest person on board (often a 14-year-old boy). In addition to his culinary duties, when the nets were being hauled in he would be lowered into the rope room to carefully coil up the incoming main rope – all one and a half miles of it.

Life on board was cramped, with little privacy and little in the way of toilet facilities; later steam drifters were built with an onboard lavatory but prior to that you 'did your business' over the side of the boat. Gerald Tungate (see Chapter 9) recalls having to use a small barrel with a rope seat around the top of it and to then empty it over the side (this was after the Second World War). When in harbour, they could only use the barrel inside the wheelhouse. If it was low tide in harbour, they stuck newspapers over the wheelhouse windows to prevent anyone looking down from the quay from seeing in! The men washed themselves in the engine room using a bucket of hot water from the boiler.

All clothing was soon covered in herring scales so the atmosphere below deck, if you could smell it through the thick tobacco smoke, was fairly pungent. Clothes could be washed with hot water from the steam geysers and then hung up to dry in the boiler room. Many Scots fishermen would take one change of clothing to East Anglia and each week they sent a bag of dirty washing home by Royal Mail to their wives or mothers. When this was clean and dry it was posted back, to be collected from one of the fish sales companies.

Crewmen slept in small bunks in the stern or in the fo'c's'le and until flock mattresses became available in the 1940s, they slept on sacks of wheat chaff, sometimes referred to as 'donkeys' breakfasts'. Gerald Tungate recalls sleeping on a sack of oat straw in 1946 when he was on the drifters. There would usually be eight bunks for ten men because there would always be two men on watch. Meals were eaten at a table next to the bunks in the stern.

The Daily Routine on a Steam Drifter

The drifters left harbour in the afternoon and steamed out to the fishing grounds, which could be anything between 30 and 60 miles out, arriving early evening. The Skipper chose his spot from experience and by watching for 'indications', such as gannets diving, the appearance of certain plankton at the surface or the presence of whales and porpoises. Casting the nets entailed attaching each net to the 'messenger' or 'bush' rope (in Scotland this rope was known as the 'leader') and tying buoys ('buffs', 'bowls' or 'pallets') to the top of the nets to keep them hanging vertically just below the surface. This process continued until all the nets (between seventy and ninety) were strung out in a straight line from the drifter. English boats, by superstition, tended to use an odd number of nets, i.e. sixty-nine or seventy-one rather than seventy. Then, with the mizzen sail hoisted, to give a little bit of steering, and with the engine at a standstill, the boat and its train of nets would simply be carried on the current, waiting for herring to rise up to feed during darkness.

Once the nets were cast, there was time for a meal and some sleep until the early hours, when the Skipper judged that it was time to take a 'look on'. If the first nets produced very few herrings, they might leave the remaining nets for a couple more hours before hauling in earnest. The messenger rope was hauled in by steam power but each net (35 yards long and 12 yards deep) had to be individually detached from it and lifted on board by hand. Every few yards of net was shaken hard to free the herrings and make them drop on to the deck (this was called 'scudding'). Depending on the size of the catch, hauling could take between four and eight hours but sometimes it could take longer. Once the herring were safely in the hold, the skipper made for port at top speed – the earliest landings and the freshest fish got the best prices in the fish market. Catch sizes could vary enormously. For all your efforts at sea, on some days you would not see a single herring, but when you did find them the catch could be anything between two crans and two hundred. There were days when the catch was so small that it was not worth the coal consumption of going back to port. In this case, the few herring in the hold would be sprinkled with salt to preserve them (these were referred to as 'overdays') and the boat would stay out for a second night.

The run for home was an opportunity to eat breakfast (usually fried herrings, up to ten per person!) and, unless there were nets that needed repairing or damaged ropes to be spliced, perhaps another hour of sleep before entering port. As soon as the Skipper had nudged the boat into a space at the crowded quay, the Mate leapt ashore to take an eighth of a cran basket of herring, a fair sample of the catch, into the fish market where the catch was auctioned. The buyer would send his cart to the vessel and the process of unloading would begin, the fish being shovelled into quarter cran baskets, using galvanised hand shovels called 'scutchers'. The baskets were hoisted out of the hold using the steam capstan and loaded on to the lorry.

While this was going on, a stream of 'runners' from local traders jumped aboard to take orders for meat, bread and vegetables, which would be quickly delivered to the boat. If fuel was low, an order was placed with a coal merchant's runner. The

Skipper might call in at the agent's office to see if any mail had arrived for the boat and to learn what catches other vessels had landed and where they had been fishing. At about 3 pm they would sail again and this routine continued, day in day out.

The Scots fishermen always observed the Sabbath, so on Saturday afternoons they would tie up in harbour and remain there till midnight on Sunday. This gave some of the crew the opportunity to sample Saturday night on the town – fine if you had all the attractions of Yarmouth and Lowestoft but if you were in the Outer Hebrides your choice of entertainment would be limited. One magistrate in Yarmouth, who had become accustomed to seeing Scots fishermen stand before him on drink-related offences, used to lecture them as he fined them: 'You earn your money like horses and spend it like asses!'

The weekend stopover could give the Driver a chance to carry out routine maintenance on the boiler and the engine but he had to stop working at midnight on Saturday. On Sundays many Scots fishermen went to a church service – the Scottish churches sent ministers down to the English ports to tend to the spiritual needs of the fishermen and other Scots workers. The less religious took the opportunity to walk, to visit the cinema or to just catch up on sleep. Day-to-day physical and spiritual needs of all fishermen were addressed by the various Fishermen's Missions, who set up facilities in many fishing ports, providing food and shelter for fishermen as well as sending mission ships out to sea with the fleet.

While the crew worked at sea, they relied on their agent in the port to collect their fish sales money, to pay tradesmen's bills, landing dues and port fees and even sometimes to send some money home to their families. If the vessel was owned by a large company, there was usually a representative in port to handle all the administration and back-up services.

Fishermen's Incomes, Early 1900s

Herring fishermen were not employed on a regular wage; instead they received a share of the boat's profits once the expenses for the season had been paid. In a successful season a crewman's share could be quite good, but in a bad season it could mean that once the expenses had been covered he took home very little.

The profit after expenses was split between the boat's owners and the crew. In England the share system reflected the fact that many boats were owned by large companies. At Lowestoft the owners took 62½ per cent, with the remainder going to the crew, and at Yarmouth owners took 55½ per cent. The crew's percentage was split into ten equal shares and these were apportioned to the crew according to their role and seniority. The Skipper would receive one and a quarter shares (Lowestoft) or one and three quarters (Yarmouth) and the cook half a share, with the remainder of the crew receiving varying amounts in between.

In Scotland, where shared ownership of boats and of the nets was common, the share system worked differently and, perhaps, more democratically. In Hopeman, for example, the profit after expenses was divided into three equal parts. The first third went to 'the boat', which meant the owner or joint owners. The second third went to 'the nets' – this was apportioned to crewmen and other individuals according to the number of nets that they owned on that boat. The final third was split into equal shares among all the crewmen (although a young learner might receive just half a share). This meant that a crewman who owned a share in the boat and owned several nets would receive income by all three of these means.

In the 1900s, Scottish herring fishermen were earning £90 in a good year and £30 to £40 in a poor year. Those who had a share in a boat or owned nets earned considerably more, sometimes as much as £300. On the East Anglian boats, it is estimated that in a good year a skipper could earn on average around £200. These figures compare favourably with the average annual earnings in 1910 for agricultural labourers of £46, and for miners of £89. There was also the perk of 'stocker' fish; once the herring nets were cast, the crewmen on some boats used hand lines to catch small quantities of whitefish such as cod, which they salted on board and sold at the end of the trip, usually for a bit of beer money.

Pay day was at the end of each season but some of the boat-owning companies used to pay crews' wives a weekly allowance or 'allotment' as a form of advance. At Winterton in Norfolk, while the fishermen were away on the Scotch Voyage, the village carrier would call once a week at the agents' offices in Yarmouth to collect the allotments for the men's families.

As steam drifters became more numerous, it became the norm for the Engineers, whose skills were in short supply, to receive a fixed weekly wage (regardless of how successful the season was). This ensured that in a poor season they did not suffer financially, like the rest of the crew.

A few fishermen, stand-ins and seasonal extras, were also paid only a fixed weekly wage but one that was much lower that the Engineer's. They could be found on both English and Scottish boats and, somewhat unfairly, no matter how hard they worked they did not share in any of the additional profits that they helped earn.

Continued Growth

The rapid take up of the steam drifter led to increased herring landings, particularly in England (Table 5). As a result of this growth, Yarmouth and Lowestoft both experienced severe congestion and struggled to cope with the growing numbers of fishing boats and the sheer volume of herrings. In order to relieve the overcrowding at Yarmouth and to accommodate all the visiting Scots fishing boats, the Borough Council developed an overflow fish quay across the River Yare at Gorleston. This had its own pickling plots, fish sales and coaling facilities, but for statistical purposes its herring landings were included in Yarmouth's.

Lowestoft was even more crowded, so in 1907 some £60,000 (including £15,000 from the government) was spent on improving harbour facilities at nearby Southwold in

Year	Quantity
1901	2,452,848
1902	3,482,736
1903	3,058,836
1904	3,198,304
1905	3,062,065
1906	3,278,289
1907	4,439,554

Table 5 English herring landings 1901–07.

order to attract Scottish boats and curers. Some of the money went towards building a large, circular, covered fish sales room which was soon nicknamed 'the Kipperdrome'. During the 1908 herring season, Southwold received forty-six landings from steam drifters, sixty-seven from sailing drifters and over 500 from smaller inshore sailing craft. A total of 14,700 cwts were landed and in 1909 this rose to 25,800 cwts (a fraction of the 1,450,000 cwts landed at Lowestoft). While Gorleston's herring quay continued to operate into the 1950s, Southwold's importance faded after the First World War and it was only small local inshore boats that continued to land herring there.

There were also changes elsewhere; in the Orkneys, the ancient port of Stromness had once been an important herring station but around 1910 the herring landings started to drop. On the opposite side of the archipelago, a new station was established on the small eastern island of Stronsay and this soon became the main centre of the herring industry on Orkney.

While this spell of growth and investment was going on in the fishing industry, in 1902 the Marine Biological Association opened a research establishment at Lowestoft. Initially its remit was to study the plaice fishery but this soon expanded to include investigating the herring populations. The 1900s also saw the introduction of ice-production factories at fishing ports. Buying fresh ice from a factory was more convenient than collecting and storing winter ice and, in 1910, Yarmouth's last remaining ice house became redundant and was subsequently used for general storage.

Between Seasons

At the end of each season, the drifters headed back to their home ports and villages. There was little time for rest as the boat and its equipment needed overhaul, repair and painting. Nets always required attention and the boiler needed servicing. If the drifter belonged to a large company it would probably have its own in-house maintenance team and staff to take care of the nets. After repair, the nets were 'tanned' (England) or 'barked' (Scotland), a treatment that helped preserve them from the effects of prolonged exposure to salt water and to jellyfish. It involved

Scottish drifters, one with an Inverness registration and the other from Banff, attract attention at the newly improved harbour at Southwold. The year is probably 1909 and the circular building is the new herring sales room, dubbed the 'Kipperdrome' by Southwold's residents.

The harbour at Buckie is here packed with steam drifters. In 1911, no fewer than fifty-nine herring drifters from Buckie worked out of Yarmouth during the East Anglian autumn fishery.

This building in Yarmouth was the last surviving ice house in the town and it became redundant in 1910. Before the development of ice-making factories, the fishing industry had to rely on winter ice, which was collected locally or brought in by ship from Norway and then stored in ice houses, insulated with straw to slow down the melting.

dipping the bundled net into a large copper or tank containing a boiling solution of 'cutch'. Cutch or catechu is an extract from acacia trees in South East Asia and is high in natural tannins. It was bought in blocks, which had to be broken up and dissolved in boiling water. The nets were dried outdoors before being folded up and packed away, ready for the next voyage. In Lowestoft you can still see some of the drying poles on the North Denes over which the nets were hung and dried. In Scottish fishing villages the nets were usually spread out to dry over dunes, hedges and railway fences.

Work on Shore

An often-quoted statistic is that for every drifter at sea during the herring season, there were ten jobs on land. Some of these jobs were to do with the boats; apart from building them (materials supplied by local timber merchants and chandlers) and constructing their engines, they required the manufacture of large quantities of nets and ropes, corks, floats and sails. Nets were frequently torn while fishing and needed repair by teams of mostly female workers. In East Anglia the large boat-owning companies had their own net warehouses with 'ransackers' (net checkers) who passed damaged nets to teams of women 'beatsters' (net repairers). In Scotland this work was generally carried out by the families of the crew.

At the harbour, a host of seasonal labourers were employed as porters, berthsmen, checkers, night watchmen and basket stewards. In the fish market there were salesmen, clerks, fish buyers, carters and lorry drivers. Local merchants delivered ice, salt and coal to the boats. On the transport side, there were additional railway workers loading boxes of herring on to special trains for despatch around the country and dockers and merchant seamen loading up ships with herrings bound for the Continent.

Processing the herring were the curers and their staff of foremen, coopers and herring gutters (more of whom later). In the smoke houses there were more gutters, as well as labourers and skilled smokers, and in the canning factories there was production line work for both men and women.

Away from the harbour, the towns' tradesmen were kept busy; butchers, bakers, grocers, overall manufacturers and hardware shops. In Yarmouth and Lowestoft the herring girls lived in lodgings, thus providing landladies with an extension to the holiday season. The fancy goods shops in these towns would see an increase in trade, especially when the migrant workers were paid off and spent their earnings (the Scots workers would always buy presents to take back to their families). Pubs did very well (fishermen were a thirsty lot!) and even the Post Office needed extra staff to cope with the despatch of parcels of kippers.

When you tot up all the economic activity generated from having several hundred fishing boats unloading in your town every day for eight weeks, the financial benefits become obvious. It was well worth putting up with a few weeks of crowded streets and general disturbance. Life soon quietened down after the end of the herring season when the industry had moved on elsewhere.

HERRING PEOPLE: Tom Bruce

Tom was the oldest of eight children and in 1902, at the age of 14, he had left school in Hopeman and obtained a job in a bank at the neighbouring village of Burghead. Although it was secure employment with long-term prospects and respectability, the money was poor. So, in 1904, in order to make a bigger contribution to the family income, he left the bank to join his father, John, at the herring fishing. He signed on as cook on the sailing drifter *Morning Star*, working alongside his father. Later that year, John Bruce bought a one-fifth share in the steam drifter *Charity* (INS 40) and they both moved to that vessel, Tom again working as cook.

Residents of the fishing villages in the Moray Firth shared a limited number of surnames and a tradition developed of giving every family a 'tee' name – a nick-name that was handed down from father to son. So, to distinguish them from other Bruces in the area, John and Tom were known locally as John 'Teen' and Tom 'Teen'.

Herring nets required regular treatment with preservatives. Here a group of Moray Firth boats are engaged in on-board 'barking' at Stornoway. Some steam drifters carried a tank of 'cutch' solution which could be heated using steam from the boat's boiler.

Net drying posts at Lowestoft. These few remaining racks on Lowestoft's North Denes are where hundreds of drift nets would once have been dried after tanning. Today they form part of the North Lowestoft Conservation Area.

Postcards depicting aspects of the herring industry were popular with visitors to the East Anglian herring ports.

HERRING PEOPLE: James Bloomfield

The Tyneside shipbuilders, Smiths Dock, established a herring fishing company in 1900 which it based at Great Yarmouth, by then the herring capital of the world. Within a few years, they were profitably operating thirty steam drifters. Their naming of boats was a bit strange in that they were all allocated numbers as names (*ONE*, *FIVE*, *ELEVEN* and so on), giving Smiths Dock Trust Company the nickname 'The Numerical Fleet'. In 1902 they appointed 32-year-old James Bloomfield as their Chief Engineer. James had come to Yarmouth from Boston, Lincolnshire, where in five years as manager of The Boston Deep Sea Fishing & Ice Company, he had turned round that company's fortunes. In 1904, Smiths Dock promoted James to General Manager.

A married man with a family, James lived in a large house near the seafront in Great Yarmouth. He was involved in various activities in the town. In 1908, he was elected to the Great Yarmouth Borough Council, where he served on the Fish Wharf and Quays Committee and was involved in running the 1909 annual conference of the National Sea Fisheries Protection Association, held in Yarmouth. Probably through this event, he met Andrew Bremner and the two hit it off well enough to have joint business dealings in 1912. James was also a director of the Great Yarmouth Fishermen's Widows & Orphans Fund, a charity established to provide financial support to the dependents of those lost at sea. This charity was funded by charging Yarmouth boat owners an annual levy.

As well as his full-time job and performing his civic duties, James also found the time and energy for an extra-curricular activity – the shared ownership of a steam drifter, the *Ocean Gift*. Her co-owner was skipper William ('Wee') Green of Winterton, Norfolk. In 1907 James and Wee each put up £400 and, with a loan from the London & Provincial Bank, they bought and equipped the *Ocean Gift* for £3,600. Over the next four years they made total profits from her of nearly £6,000 (at least half a million pounds at today's values).

Timelines Chapter 1

1900 Smiths Dock Trust Co. is formed at Yarmouth

1900 The Dunbar fishing laboratory moves to Torry, Aberdeen

1901 The first experimental motor-powered drifter is built at Lowestoft

1901 An international conference on North Sea fishing is held at Christiania (Oslo)

1902 Lowestoft Fisheries Research Laboratory is founded

1902 At Wick, the skipper of the *Strathnaver* is stoned for landing trawled herrings

1903 Richard Irvin Ltd begins producing 'Tyne Brand' tinned herring at North Shields

1904 James Bloomfield is appointed manager of Smiths Dock Trust Co.

1907 Severe congestion in Lowestoft and Yarmouth harbours during the herring season

1907 Southwold (Suffolk) harbour is redeveloped in order to attract the herring industry

1907 James Bloomfield and 'Wee' Green buy the steam drifter *Ocean Gift*

1908 At Southwold 300 Scottish drifters land herring which are processed by Scots girls

1908 The first herring reduction factory in Europe is opened in Norway

1908 The Cran Measures Act passed and the cran replaces the 'Last' in England

1910 The final remaining ice house in Yarmouth closes, replaced by factory-produced ice.

CHAPTER 2

1910 to 1914 –
The Glory Years

As the herring industry moved into the second decade of the twentieth century, it was experiencing still more growth and expansion. More steam drifters were being built, more herrings were being caught and, importantly, more were being exported. On paper, Germany was still the biggest buyer of British herring, ahead of Russia. However, the Scottish Fishery Board believed that some 40 per cent of the herrings sold to Germany were subsequently resold to Russia. Many of these re-exported herring were from the 'Klondyke' trade, unpacked on arrival in Germany and then gutted, salted and barrelled there. One estimate was that by 1914 some 70 per cent of British herring exports were ending up in Russia.

More people than ever were working in the herring industry. In an industry that employed many temporary and seasonal staff, it is not easy to find accurate numbers of permanent staff. However, in 1912 official statistics indicated that in Scotland alone there were 21,700 fishermen, 2,100 coopers and 10,800 gutters who all depended on the herring industry for their livelihood. In England there were probably a further 10,000 full-time herring fishermen.

1911 - A New Company Enters the Industry

Had the herring industry reached its peak? In Great Yarmouth, James Bloomfield was confident that it hadn't. His co-ownership of the drifter *Ocean Gift* had given him an appetite for being his own boss. On 28 March 1911, having found some influential backers, he registered a new company, Bloomfield's Limited. As soon as he had worked his notice at Smiths Dock Trust Co., he was installed as Managing Director of the new company, on a salary of £800 per annum (around £280,000 at today's values). The company established its head office, Ocean House, close to the Fish Quay in Great Yarmouth.

The flotation of the new company was 50,000 shares at £1 each, giving the company a modern-day value of around £18 million. The shares were oversubscribed and Bloomfield's Limited enjoyed the luxury of picking whom they would allow to buy their shares. The prospectus for the new company listed

The hub of Bloomfield's operations was situated on South Denes Road, Great Yarmouth, facing the Quay. Under the auspices of Lady Savile-Crossley, the building became for a short while a War Hospital in 1914-15. If you look closely, you can see net-drying poles on the bare land to the right and also on some of the flat roofs.

herring fishing and vessel-owning as its main aims but it also included a host of other fishing-related activities, including curing, ice manufacture, fish selling, sail making and salt trading. James Bloomfield clearly intended from the very start to be a major player in the fishing industry and was confident that money was to be made in all these different sectors within the herring industry.

The largest shareholder was not James himself but the chairman, Charles Sydney Goldman, the Member of Parliament for Falmouth and Penryn in Cornwall. The other Bloomfield's directors came from the City of London and from the manufacturing industries of Yarmouth and Aberdeen. Of the many ordinary shareholders some, as you might expect, had links with the sea and these included shipbuilders, skippers and engineers. There were many more that had no maritime links and the professions in this list included: tobacco planter, bank manager, draper, theatre proprietor, Clerk in Holy Orders, East India merchant, cotton manufacturer, pottery manufacturer and farmer. The herring industry was clearly seen as a good area in which to invest.

Financial Success at Ocean House

In the summer of 1911, within weeks of starting the company, James Bloomfield sent four steam drifters on the Scotch Voyage and by the autumn Home Fishing he had eight in operation. Following the example set with *Ocean Gift*, the names of most Bloomfield's vessels carried that same prefix, giving the company the nick-name 'the Ocean Fleet'. Two of the eight boats were part-owned by their skippers, one of whom was James' earlier business partner, Wee Green. The shared-ownership scheme was a good incentive and made James popular with his skippers. At the end of 1911 Bloomfield's Limited had made a net profit of £3,600 and paid its shareholders a dividend of 10 per cent (making James also popular with his shareholders!). In early 1912 the company issued another 12,500 shares and these too were quickly snapped up. By the end of that year, the fleet had grown to ten drifters owned outright and five part-owned and, with branch offices from Lerwick to Londonderry and Lowestoft, they doubled the previous year's profit. In the light of these results, the Board agreed in early 1913 to increase James' annual salary to £1,000.

The 1911 Yarmouth Herring Season – Records Tumble

In the 1911 herring season, a total of 990 fishing boats were based at Yarmouth, each one paying between £1 and 30 shillings in mooring fees for the season. Lowestoft was host to another 600 vessels. Of those at Yarmouth, 357 were East Anglian vessels. The remainder (nearly two thirds of the total) came from other parts of the country, including Yorkshire, Northumberland, the Isle of Man and Plymouth. The majority (549), however, came from Scotland of which twenty-two were from Hopeman, fifty-nine from Buckie and seventy-one from Peterhead. These numbers are a further indication of just how much money the small Scottish fishing communities had invested in their boats and how dependent they were on the herring industry. Of the Yarmouth-registered boats, most were owned by limited companies but some, such as those with connections to the villages of Winterton and Caister-on-Sea, were owned by individuals and by families. Among the large companies fishing out of Yarmouth that season were Smiths Dock, with thirty-two steam drifters; Richard Irvin of North Shields with twenty-eight; and W. Leask of Aberdeen with twenty-three. Other fishing companies represented were The Girdleness Steam Herring Drifting Co. Ltd (of Aberdeen), Horatio Fenner Ltd (Yarmouth) and The Britannia Fishing Co. Ltd (Lowestoft).

The 1911 East Anglian landings were a record 523,000 crans at Yarmouth and 435,000 crans at Lowestoft. These figures prompted speculation as to whether such huge landings could ever be repeated. That year the first cargo of herrings sailed from Yarmouth for northern Europe on 7 October and from then till mid-December there was an average of two ships leaving each day carrying Klondyked or salted herrings. Incoming vessels at Yarmouth and all the larger herring ports and stations would be carrying cargoes of empty barrels, salt and coal.

With increasing prosperity in the herring trade, it now became quite the thing for drifter owners or skippers to have a model built of their vessel or, more commonly, to have a picture painted of the boat. A small number of artists were able to make a living by moving from port to port, producing naïve paintings of fishing vessels which they sold for a few shillings. These artists are referred to nowadays as 'pierhead painters' and the best-known of them are probably George Race and E. G. Tench. The images had to be painted very quickly (sometimes overnight) in order to catch the buyer while he still had money in his pocket or before he went to sea again. Today, these paintings can sometimes fetch four-figure sums in art auctions.

Month	No. of Sailings
October	39
November	63
December	39

Table 6 Herring export departures from Yarmouth, 1911.

Seasonal Riches

For three months of each year, the large herring ports such as Lerwick, Wick, Buckie, Fraserburgh, Peterhead, North Shields, Yarmouth and Lowestoft became boom towns. Imagine the services needed on shore to support hundreds of fishing boats, each with up to ten crewmen on board. As well as the fishermen, the visiting workforce included the curers, hundreds of coopers and thousands of gutters. The streets around the quays and docks were bustling with lorries, carts, herring workers and foreign buyers. The harbours were packed with drifters jostling for an unloading space at the quay and close by there were cargo vessels loading up with Klondyke herrings for Europe. There was constant noise from the steam whistles of the drifters and shouting from the fish auctions and the hammering of lids on to fish boxes. English, Scottish and Gaelic accents mixed occasionally with German, Dutch and Russian voices. A permanent pall of coal smoke from the steam drifters, mixed with wood smoke from the smokehouses, hung over the towns, along with the smell of kippers from smokehouses.

In Yarmouth the Fish Quay was a mile away from the town centre and businesses would open up temporary offices there to deal with the additional business generated during the herring season. In 1911, for instance, organisations running seasonal offices at the Fish Quay included the Post Office, the Salt Union, four banks and five railway companies.

Portrait of a Lowestoft fishing vessel. This image of *Galilean* LT 1128 by an anonymous pier head painter was one of many similar paintings that were commissioned by drifter skippers and owners in the early 1900s.

Not a Good Year for Everybody

While some in the Yarmouth fishing community were celebrating their good fortune in 1911, others were not faring so well. The Fishermen's Widows and Orphans Fund was kept busy with applications for relief, especially in the autumn when two vessels were lost (see Appendix 1). The Fund's administrators did their best to assess each application on its merits and to allocate fairly the limited funds they had at their disposal.

Every line in Appendix 1 tells a story of tragedy and of families plunged deep into poverty. At a time when the only state benefit was the newly-introduced old age pension (5 shillings a week), the small sums paid out by the Fund (7 shillings a week for a widow and 2 shillings for a dependent child) were often all that stood between the bereaved and the workhouse. These benefits were payable normally for two years and could only be extended after the directors had re-examined the case. In instances where the deceased fisherman was unmarried, the Fund would award a one-off grant of £10 to dependent parents. While on the subject of the workhouse, in October of 1911 one of the inmates on the Mental Ward of Yarmouth Workhouse is recorded as 'Robert Young, a Scotch fisherman'. His confinement there is, perhaps, another indication of the difficulties of working in the herring industry.

During the spring of 1913, there was industrial unrest in northern Scotland when numbers of hired men (seasonal deckhands who received a fixed wage rather than a share) became militant. The weekly trade newspaper, *Fishing News*, carried reports of strikes at Buckie, Stornoway and Hopeman, where the hired men all wanted a fairer system of pay, one under which they too could benefit from the boom in the herring industry.

Now in a sad state of repair and boarded up, the Refreshment Rooms was once the hub of Yarmouth's Fish Quay. In its various bars (there was a Skippers' Room, a Crew Members' Room and a Mates' Room) business deals were hammered out and crews were engaged.

This busy quayside scene from the early 1900s gives an idea of the number of jobs that the herring season brought with it. This is Yarmouth but there were similar scenes at all herring ports.

The 1911 National Insurance Act

In the buoyant industry of 1911, the herring fishermen were earning good money, buying their own homes and investing in new boats. That year, however, they lost out on one important new development which at the time probably did not seem too serious to them. In 1911 the Liberal Government's National Insurance Act became the first tentative step towards universal social and health care. It was generally welcomed by working people who, at this stage, were its only beneficiaries. The Act was in two sections; Part 1 related to health insurance and entitled all who paid an insurance 'stamp' (7d per week) to free healthcare and sickness benefit. Part 2 of the Act introduced an entitlement to unemployment benefit of 7s (35p) per week in return for an additional contribution of 2½d per week.

Deep in the small print, the exclusions from the scheme included 'share fishermen', meaning that most herring fishermen would not qualify for unemployment benefit but, while they were all earning good money during a boom, no-one was too worried about this – after all, they were saving themselves 2½d per week.

Under Part 1 of the Act, you registered with the doctor nearest to your home. But what about when you were working away from home for weeks on end? In 1911 the newly-formed Great Yarmouth Insurance Committee were very concerned about how they and the Yarmouth medical services could cope with their new responsibility to provide healthcare to the autumn influx of thousands of Scottish herring workers. The Committee's clerks put in extra hours, working sometimes to midnight, to issue temporary green vouchers to the thousands of Scots workers and the system coped.

Trawlers *v.* Drifters – An International Conference

The dispute over herring trawling was still rumbling on. Despite having once worked for a trawling company, James Bloomfield was not going to sit back and permit a possible threat to the herring trade (and to his new company) so he entered the argument. In 1912 he founded the National Herring Fisheries Protection Association (NHFPA), to combat the destruction of English and Scottish herring stocks by trawl fishing. In an industry which then consisted of many local herring trade bodies, the NHFPA was the first truly national organisation. Another leading member of the NHFPA was George Slater, a herring merchant and curer from Hopeman who would have been well known to John and Tom Bruce. In October 1912, James organised a national meeting at Great Yarmouth to gather views about the trawling of herring in the North Sea. The meeting sent a deputation (which included three representatives from Hopeman) to Whitehall to speak to the President of the Board of Agriculture and Fisheries.

Three months later, a conference was held in London, attended by representatives of both drifting and trawling. The trawling interests vigorously defended themselves, maintaining that they were not damaging herring stocks and that most of the herring that they caught were of saleable quality. The NHFPA countered this with recent evidence of a trawling trip from Hull on which seven baskets of saleable herring were landed, while twenty-four baskets of dead, immature herring had to be thrown overboard as being too small to take to market.

The climax of the campaign against herring trawling was an international conference held in London in early 1914 at Fishmongers Hall (a very apt venue!). At this event, the English contingent comprised four delegates from Yarmouth (including James Bloomfield), three from Lowestoft (including Frederick Spashett of Small & Co.) and one from North Shields. Scotland sent seven delegates, France two, Germany two, Russia two and Holland four. The speakers all stressed the importance of the herring industry to the economy of their respective nations and the damage and waste caused by herring trawling. After a rousing speech from James Bloomfield, the conference resolved that all delegates should call on their governments to protect the herring industry from the damage caused by trawl nets and to persuade the North Seas Convention to ban the trawl nets currently being used for herring fishing.

The British Prime Minister appointed a Parliamentary committee to inquire into the whole question of herring trawling – the government could hardly ignore the problem as the herring industry at this time was a hugely important one in terms of exports. Unfortunately, war broke out before the Committee reached any conclusion.

The Herring Fishing Branding Act, 1913

In October 1913, the English herring industry (with considerable urging from James Bloomfield) moved a step closer to matching Scotland's quality controls when the Herring Fishing Branding Act was passed by Parliament. This Act would have led to a branding system for barrels of English salted herring similar to the long-established Crown Brand in Scotland. The Act came into force in 1914 but only 250 barrels were branded before war intervened.

1913: The Best Year Ever

The British herring industry was still going from strength to strength, as illustrated by the annual landings at Great Yarmouth (see below).
The run of good seasons had persuaded even more fishermen to invest in steam and by 1913 there were 1,800 steam drifters in operation, some two thirds of which were Scottish-owned.

The 1913 summer herring fishing in Scotland had produced a record total of 1,324,000 crans. In the autumn a fleet of over 1,100 Scottish drifters (predominantly steam but including 100 motor and 200 sail) headed south for the East Anglian season. The combined English and Scots fleet once again broke all records, landing

Season	Crans
1908	476,730
1909	483,000
1910	381,600
1911	523,300
1912	683,600

Table 7 Herring landings at Yarmouth, 1908–12.

824,000 crans at Yarmouth and 536,400 crans at Lowestoft. At Lowestoft on Monday 12 October, no fewer than thirty-three drifters each landed catches of over 200 crans – an amazing volume of herrings for the trade to cope with.

The total herring catch in England and Wales for 1913 was 7.3 million hundredweight (two million crans), of which some 70 per cent was landed in East Anglia and 20 per cent in Yorkshire and the North-East. Importantly, the markets and the curers were able to cope with the huge numbers and nearly all the herring caught and cured were sold to Europe (See Appendix 2). Total exports of British cured herring in 1913 were nearly ten million hundredweight (half a million tons), with a value of £5.9 million (£500 million at today's values).

With all those herring to export, it is no surprise that the number of sailings of cargoes of herring from Yarmouth was higher than ever. The first shipment was despatched on 10 October and the last on 11 March 1914. On one day alone, no fewer than ten vessels carrying herrings left Yarmouth for the Continent and the average throughout the season was four sailings a day.

Drifter Finances, 1913

Why were people so keen to own steam drifters and invest in the herring industry? The table below gives us a clue. It details the income and expenditure for one of the Bloomfield's drifters. This vessel was built in 1911 at a cost of around £3,000

Month	No. of Sailings
October 1912	50
November 1912	126
December 1912	53
January 1913	25
February 1913	11
March 1913	1

Table 8 Herring export departures from Yarmouth, 1913.

and with an expected asset life of 20 years (she was actually still working in 1939 – some steam drifters had even longer working lives.)

If the gross profit of £1,472 could be repeated for a further two years, the original capital investment in the drifter would be paid off – a return on capital that many businessmen could only dream of. This table does not show the full picture because the total voyage time shown here covers only eight months. If they wanted, East Anglian boatowners could (and they often did!) squeeze in a couple of months'

Trading Account for 1913, *Ocean Angler*			
	Scottish Voyage (121 days)	Home Voyage (98 days)	Total (229 days)
Expenditure	£	£	£
Commission	56	97	153
Landing/Harbour Dues	14	28	42
Coal	250	122	372
Salt	6	39	45
Provisions	81	65	146
Stores	44	26	70
Water	5	8	13
Sundries	31	58	89
Crew Wages	350	585	935
Total Expenditure	837	1,028	1,865
Income			
Herring Sales	1,390	1,947	3,337
Gross Profit	553	919	1,472

Table 9 Steam drifter finances, 1913.

winter fishing in the South-West for herring or mackerel before preparing the boat for the next Scotch Voyage.

The wages figure (or, rather, the crew's percentage share) of £935 gives average earnings of £93 per man. In reality, of course, the complicated share system would produce differing figures for each crewman. After wages the largest item of expenditure for *Ocean Angler* was coal, which became more expensive the further away you were from the coalfields (and you could not get much further than the Shetlands). One of the first things that James Bloomfield had done on forming his own company in 1911 was to buy a coal hulk, fill it with coal in Durham and tow it to Lerwick harbour to ensure that his vessels had access to cheaper coal. James regularly campaigned to get the railway companies to reduce their charges for transporting coal to the East Anglian ports. In Yarmouth and Lowestoft coal cost considerably more than in Hull, Grimsby, North Shields or Fleetwood, these being ports which were situated near to coalfields.

HERRING PEOPLE: James Bloomfield

In 1911, James Bloomfield had stepped down from Yarmouth Borough Council and from the Widows and Orphans Fund. Despite being busy running his own business and playing a very active role in the NHFPA, he was able to capitalise on his growing reputation. In 1912, the City & Peterhead Drifting Company hired him as a consultant to advise them on the management of their six drifters. In the same year he accepted a directorship of the Total Loss Mutual Steamship Insurance Company, which had its headquarters in Sunderland.

By now James enjoyed the benefit of a company car, and of having a telephone in his house which was paid for by the company. As Bloomfield's Limited's profits grew, he was able to buy more shares in the company.

HERRING PEOPLE: Tom Bruce

In Hopeman Tom Bruce had left his father's drifter, the *Charity*, to join another Hopeman steam drifter. He passed his Board of Trade competency examination in April 1911, thus gaining his Skipper's 'ticket' (certificate number 11135). Meanwhile, in 1912 his father, John, increased his share in the *Charity* to one third and he became her Skipper.

In the increasingly prosperous world of the herring industry Tom felt financially secure and in 1913 he married a Hopeman girl, Jessie McPherson. She, too, was from a fishing family – this was Hopeman, after all – and she had worked as a herring gutter in the Orkneys and at Yarmouth.

The 1913 herring bonanza convinced Tom and two Hopeman friends, Alec Lawson and Willie Main, that they should join together and buy their own steam drifter. They raised the money from savings and by taking out loans, including one from Lachie Mackintosh, a local merchant and investor, and ordered a vessel to be built at nearby Lossiemouth. They may also have borrowed from Richard Irvin & Sons, who were known to lend money for building drifters to fishermen in the Hopeman area. Their new wood-built drifter was delivered in early 1914. They named her *Admiration*, with the port registration number INS 99. In the early summer of 1914 they began fishing with her in the Shetlands.

HERRING PEOPLE: Andrew Bremner

Andrew Bremner was busy during the Scottish summer of 1913, simultaneously running curing yards at Stornoway, Wick and Lerwick. He certainly reaped the benefits of that year's herring boom because in October 1913 he bought a new home. He sold 8 Breadalbane Crescent in Wick and moved to the Keiss estate, a few miles north. This estate comprised Keiss Castle, an impressive mansion built

James Bloomfield (centre) and his wife appear to be enjoying a social event with the skippers and mates of the company vessels. This was probably taken just before the First World War.

Every six months the owner of any merchant vessel, including all fishing boats, had to submit to the Registrar General of Shipping and Seamen a list of all the crew, their ages and when they joined the vessel. This list (summer 1914) was the last to be completed for the *Charity* of Hopeman as she was requisitioned for naval duties a few weeks later and lost in 1915.

Skipper Tom Bruce. Tom Bruce's daughter said that he always smoked a pipe and that probably the photographer had offered him the cigarette he is holding.

in the Scots Baronial style, and nearly 4,000 acres of land tenanted by farmers. So, while still running his business in Wick, he began the life of a country gentleman.

Timelines Chapter 2

1911 National Insurance Act excludes share fishermen from unemployment benefit

1911 Bloomfield's Limited is founded in Great Yarmouth

1911 990 vessels fish out of Yarmouth in the autumn herring season

1913 Total herring landings in Scotland are over 1.3 million crans

1913 The total number of steam drifters in operation reaches 1,800

1913 The Herring Fishing Branding Act is passed

1913 1,163 vessels land 824,213 crans of herring at Yarmouth

1913 536,400 crans landed at Lowestoft

1913 Tom Bruce of Hopeman buys a third share in a new steam drifter

CHAPTER 3

International Dealings

The International Scene

The herring trade had always been an international business. British smoked herrings had been exported to Mediterranean countries for centuries and now there was the far larger trade in British salted herrings to northern Europe and to the United States. In the early 1900s the United Kingdom, especially Scotland, was producing the finest quality salted herrings in Europe, and in sufficient quantities to dominate the world herring trade.

Other nations involved in herring fishing and herring trading were Norway, Sweden, Holland, Iceland, Belgium and France. On the North Sea, British herring fishermen would fish alongside Dutchmen from Scheveningen and Ijmuiden and Frenchmen from Dieppe, Boulogne and Fécamp. Occasionally, if prices were poor in Yarmouth, British boats headed east and sold their catches at one of the Dutch ports. For the same reason, foreign vessels sometimes landed their herring catch at British ports – in 1913, for example, Norwegian fishing boats landed 53,000 tons of herring at United Kingdom ports. Buyers from Holland joined those from Russia and Germany in attending the various herring seasons around the British coast.

In a spirit of international co-operation, the North Sea fishing nations held a series of conferences to discuss fishing and fish stocks. Scientists from these countries first met at Stockholm in 1899, and then at Christiania (now Oslo) in 1901. The fishing nations were keen to work together regarding a shared natural resource. This spirit of co-operation included the sharing of ideas, new methods and processes.

In 1904, James Bloomfield became involved in a research and training project when the Canadian Department of Marine and Fisheries chartered the Smiths Dock steam drifter *Thirty Three*. This vessel was taken by a Scottish crew across the Atlantic to Nova Scotia, where she was used on experimental herring fishing and for training Canadians in catching and curing herring. This venture may have inspired James to become involved in later international schemes.

Above left: European herring buyers. The German and Russian 'middle men' are informing the industry that they will be buying herring in northern Scotland in the summer of 1913 (at that time Stettin, now in Poland, was in Germany).

Above right: The Mediterranean trade. In 1913 the export trade in smoked herring from Yarmouth and Lowestoft to the countries of southern Europe was an important part of the herring industry. The smoked herring were carried by cargo liners which could only use deep-water ports such as Hull and Liverpool.

Trade with Russia

Scotland had a long history of selling herrings to Russia. In the early nineteenth century James Miller, a herring curer and merchant from Leith, set up a base in St Petersburg. His son William succeeded him and after sixteen years as Honorary British Consul in St Petersburg, William came home with a fortune which enabled his descendants to build Manderston House, a stately home in the Scottish Borders.

Selling salted herring to Russia in the early 1900s involved dealing with Baltic-based middle men and merchants who sent buying agents to British herring stations, to inspect the barrelled herring and to place orders. In their quest for good quality herrings, some Russian buyers would insist on a barrel from each batch being opened so that they could take a bite out of one of the fish, checking the flavour of both the fish and of the pickling brine. During the course of any season, whether in Lerwick or Lowestoft, there would be regular visits by schooners and tramp

This curer, Donaldson, has abandoned the official Crown
Brand mark and is using just his name as a brand for his salted
herrings. Can you guess from the logo which country they were
destined for?

steamers, loading up with barrels of herring and then departing for St Petersburg
and other Baltic ports. The Fishery Board for Scotland was so keen to maintain the
good reputation of Scottish herring that it sent inspectors out to the ports of entry
to check that barrels arriving from Scotland had not leaked or been damaged in
transit.

Andrew Bremner of Wick had considerable experience of exporting herring
and of the doing business with the Russians. They were good customers and
he made a lot of money by selling to them. He had regular dealings, through
the Scottish Herring Import & Export Co. Ltd, with the St Petersburg herring
trading company J. Wolkoff. Selling in St Petersburg was not a straightforward
process because you had to negotiate the Russian bureaucracy and also pay
some hefty fees and taxes. As an example, Andrew sold Wolkoff 200 barrels of
large Prima Matjes for 11,200 roubles. By the time he had paid the import duty
of 1,400 roubles and miscellaneous charges of 1,900 roubles he received just
7,900 roubles. These miscellaneous charges included freightage (to Cronstadt,
the main port for St Petersburg), handling charges, storage costs, town
dues, administrative charges, fire insurance, harbour dues and commission.
Fortunately for Andrew, even after all these expenses there was still a profit to
be made.

Bloomfield's Begin to Forge International Links

Through international trade and conferences, a grapevine developed within the
international fishing communities and certain groups and individuals gained
reputations for their expertise and innovation. In 1912 the still brand-new company
of Bloomfield's Limited received a business proposition which could have resulted in
a very long-distance partnership. The approach came from a new fishing company,
the Trawling & Direct Fish Supply Co. of Australia, whose directors included the
explorer Sir Ernest Shackleton. Bloomfield's chairman, Charles Goldman, had
several meetings with the directors in London. The Australian company wanted
to use Bloomfield's expertise to help them develop fisheries in the Southern
Hemisphere and they wanted James Bloomfield to spend several months of each
year in Australia. The proposed partnership collapsed because Bloomfield's could

not quickly raise the £25,000 which was to be their half share in the enterprise. However, as one door closed, another opened.

In November 1912, James Bloomfield told his fellow directors that he had been involved in some negotiations with a Russian called Vladimir de Sivers. James had in fact already travelled to Berlin to meet de Sivers and an admiral of the Russian navy. The board agreed that James should proceed with the negotiations. In January 1913 de Sivers (sometimes known as 'von Sivers' or 'de Sievers') came to London with a Captain Spahde for a meeting at the Hotel Metropole with James and Charles Goldman. It was a successful meeting and both sides decided to work towards a possible partnership.

HERRING PEOPLE: The Russians

De Sivers is described in the Bloomfield's records as an 'Equerry of the Imperial Russian Court'; his home was on Galernaia, one of the grandest streets in St Petersburg. According to modern Russian sources, his family were members of the Russian aristocracy. In addition to his Court duties he was also an entrepreneur, owning two steam trawlers and several other vessels.

Captain Carle J. Spahde was a resident of Archangel and was introduced as a Russian fishing expert. By means of statistical research he had shown that while the Barents Sea and the White Sea contained large fish stocks, 80 per cent of the fish landed at the Russian ports of Murmansk and Archangel had been caught by Norwegian fishermen.

To the Russian government, keen to reduce their reliance on foreign fish imports, it was clear that their fishing industry and its methods required improvement and modernisation. Thus the reason for approaching Bloomfield's Limited was to draw on their expertise of running a modern fishing fleet. To assist in restructuring the Russian fishing industry, the Russian government had granted de Sivers substantial concessions which included remission of import duty on fish.

Advice From Scotland

On 12 March 1913, James Bloomfield updated his directors on the progress with de Sivers. Andrew Bremner, the herring curer, was invited to this meeting so that he could share his considerable experience of the Russian herring trade. He explained how in his time the Russian duty on imported herrings had doubled and he told the Board that he was confident that joining this new venture would be in the best interests of Bloomfield's Limited. (This was fortunate because at the January meeting in London some six weeks previously, a Memorandum of Agreement had been signed which had practically committed Bloomfield's to joining the venture).

The board minutes do not mention that while he was not a director, Andrew Bremner currently held the largest single shareholding in Bloomfield's Limited. In 1912 Bloomfield's had bought two steam drifters from him and instead of cash he was given 7,000 shares in the company, making him an extremely influential member of Bloomfield's Limited. (By 1914, his holding had grown to over 11,000 shares.)

A Business Trip to St Petersburg

In late March, James Bloomfield and Charles Goldman made the long journey to St Petersburg, where they met de Sivers, Spahde and the other Russian partners (see Appendix 3). While the names mean little, the list of Russian shareholders includes some very impressive military titles. (If you were working in Russian waters, it would be wise to have the Russian Navy on your side.)

The two sides formulated a joint working agreement with the stated objective of: 'Fishing, trading in fish and the exploitation of the riches of the sea of any kind whatever under the Russian flag.' The first stage of the venture would be to modernise and expand the Russian fishing fleet in the Barents Sea and the White Sea and to establish trading bases in the ports of Murmansk and Archangel.

The new syndicate (partnership) was christened 'The Russian Northern Maritime Industries', with de Sivers as chairman and James Bloomfield and Charles Goldman among the directors. The total capital for the venture was 330,000 roubles, of which Bloomfield's would provide one third. In return for this investment, Bloomfield's were to receive 4,000 roubles a year for providing advice and expertise, 7 per cent annual return on their capital and 5 per cent of the annual net profits. James Bloomfield was to 'devote his personal experience to the enterprise' – this included a responsibility to purchase vessels and equipment. The aim was that the enterprise would run for an initial period of ten years.

In May of 1913 Bloomfield's paid 110,000 roubles (£11,770 at 1913 exchange rates) into a St Petersburg bank. Using the Average Earnings Index, this sum would today equate to around four and a half million pounds.

What Were the Benefits of Joining the Russian Venture?

There are several unanswered questions surrounding this venture. On the face of it, the financial arrangements would have provided Bloomfield's with an attractive but unremarkable return in exchange for their capital and expertise. On this level, and given the uncertain political situation in Russia and Eastern Europe, it is difficult to understand why Bloomfield's were so keen to join this venture. The White Sea was remote and there were some 2,000 nautical miles between Great Yarmouth and the port of Archangel. This would have made it difficult for anyone based in England to monitor or have any control over activities out there.

Was the attraction for Bloomfield's and for Andrew Bremner really the publicly-stated objective of developing a fishing industry in Russia or were they looking for a means of avoiding of some of the hefty duties and fees being imposed on the salted herring that they were exporting to Russia? Also, just after he had first met James Bloomfield in November 1912, why did de Sivers buy 2,000 shares in Bloomfield's Limited? Another little mystery is that after James Bloomfield's trips abroad, a new name appeared on Bloomfield's register of shareholders, a Madame Jeanne Alexandrine Marie Manuel, a widow living on the Avenue d'Alma, Paris, who bought 1,330 shares.

The formation of the Russian Northern Maritime Industries received no mention in the Yarmouth press or in the UK fishing trade journals, so Bloomfield's Limited appeared to be keeping it 'under wraps'. Why, one wonders?

The Pesha

James Bloomfield had been put in charge of buying vessels and equipment for the new venture and he wasted no time; that April he bought a Hull trawler, the *Melbourne* (built 1892), for £875 and spent £800 in quickly fitting her out with driftnets and equipment for herring fishing. (Why buy a trawler instead of a drifter? Probably because trawlers had larger coal bunkers and were, therefore, better suited to sailing very long distances.) The *Melbourne* was renamed the *Pesha* and a skeleton crew from Yarmouth were signed up to work for three months for 'Charles Spahde, Fishing Boat Manager of Saint Petersburg'. The *Pesha*'s skipper was James Bloomfield's close associate Wee Green of Winterton. Wee was to receive a wage of £7 per week (£2,000 at today's values) and the other men around half that sum. According to the contracts they signed, their work would be 'on the North Sea, the Arctic Ocean and on any inlets thereof'.

The *Pesha*'s assignment was to find herring in the White Sea and to teach Russian fishermen how to use driftnets. In late May, the *Pesha* left England with Captain Spahde on board and sailed northwards. She arrived at Archangel in early June where Spahde disembarked and they took on board a number of Russian seamen. They began fishing on 13 June some 40 miles west of Archangel and then tried other locations in the White Sea, but with no success. They appear to have been particularly unlucky because Russian fishery statistics for 1913 record that some 77,000 hundredweight of herring were caught in the White Sea. After a week, Wee took the *Pesha* out into the Barents Sea and worked westwards towards Murmansk, fishing all the time but again not finding any herring.

Reaching the limit of Russian waters on 27 June, Wee followed James's back-up plan and headed for Shetland, where, for training purposes, they could be guaranteed plenty of herring. In September the *Pesha* arrived back at Yarmouth

This remarkable document is the contract between Wee Green and Captain Spahde of the Russian Venture. The fishermen signed up to work for three months to prospect for herrings in the White Sea and Barents Sea and to train Russian seamen in the techniques of drift-netting.

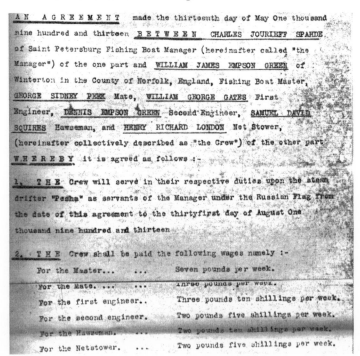

AN AGREEMENT made the thirteenth day of May One thousand nine hundred and thirteen BETWEEN CHARLES JOURIEFF SPAHDE of Saint Petersburg Fishing Boat Manager (hereinafter called "the Manager") of the one part and WILLIAM JAMES EMPSON GREEN of Winterton in the County of Norfolk, England, Fishing Boat Master, GEORGE SIDNEY PEEK Mate, WILLIAM GEORGE GATES First Engineer, DENNIS EMPSON GREEN Second Engineer, SAMUEL DAVID SQUIRES Hawseman, and HENRY RICHARD LONDON Net Stower, (hereinafter collectively described as "the Crew") of the other part WHEREBY it is agreed as follows :-

1. THE Crew will serve in their respective duties upon the steam drifter "Pesha" as servants of the Manager under the Russian Flag from the date of this agreement to the thirtyfirst day of August One thousand nine hundred and thirteen

2. THE Crew shall be paid the following wages namely :-

For the Master... ...	Seven pounds per week.
For the Mate.	Three pounds per week.
For the first engineer..	Three pounds ten shillings per week.
For the second engineer.	Two pounds five shillings per week.
For the Hawseman. ...	Two pounds ten shillings per week.
For the Netstower. ...	Two pounds five shillings per week.

from the Scotch Voyage and the Russian seamen travelled home. The arrival of the *Pesha* was accompanied by a newspaper article in Yarmouth which described the voyage as having been for instructional purposes (the partnership with the Russians was not mentioned). James registered the *Pesha* as part of the Bloomfield fleet and renamed her *Ocean Comrade* - in time for her to take part in the 1913 East Anglian 'Home Fishing'.

Skilled Artisans for the Russian Venture

A note in the Bloomfield archives says that Captain Spahde had asked to have a British man working with him in Archangel. Andrew Bremner contacted James Bloomfield, saying that he had found a Scottish couple, a cooper and his wife, who were willing to go to Russia. Their role would be to train Russian workers in gutting and curing herrings.

Financial records show that Bloomfield's paid the fares to Archangel for 'Sinclair and wife' – undoubtedly the cooper and his wife – and also that their wages were paid for May, June and July. Unfortunately, no other records have come to light of what must have been a fascinating experience for this couple.

Left Arkangel Friday 13th June for fishing grounds. Course
West 44 miles from pilot boat. No fish.

Steamed 14th 130 miles N.W. half W.

Anchored on Sunday 15th as Russians said they did not work on
Sunday as it was some sort of a holiday in their country till
Monday 2.30 Arsmovaia Bay 16th sailed 4 miles East then shot here
at 4 p.m. No fish.

Tuesday, steamed 100 miles S.E. then 15 miles South. Shot between
Solvetski Island and mainland - tide 1½ knots.

See nets coming in bunch. Started to haul at 10.30 p.m. Never
got only one net alright. Found the rope had catched a rock on
the bottom though we could not get less than 15 fathoms of water
17 when shooting.

We went back to Arkangel and I told Spahde that I did not think
there was any herring in the White Sea as the water was as brown
as a berry and no good, nothing only white jelly. I also told him
when he was on board as we were going to Arkangel at first.

He said there must be some, else how would they catch them. I
told him not with this class of net, not at least at this time
of year. I told him I thought the best thing to do was to take
300 barrels out to the hulk and try the Murmansk coast as I
thought there was better prospects there and he agreed. Then I
told him if we had a try there for a week and could get nothing
if I was him I would send us on to Lerwick to see if we could
clear ourselves there. And this he said he had thought on before
but did not like to mention it, and we left there as soon as we
had coaled on 20th June.

Wee Green's report. When Wee returned to Yarmouth from Russia, this report of the voyage was typed up from his log.

Slow Commercial Progress in Russia

Nearly a year after the formation of the syndicate, optimism was fast disappearing. De Sivers wrote to James Bloomfield in November 1913, expressing regret that he was unable to furnish any information on the trading of Russian Northern Maritime Industries as he had not heard from Captain Spahde for a considerable time. It is no surprise to learn that the first year of trading produced 'disappointing results'.

By now alarm bells were ringing in Yarmouth. De Sivers came to London in March 1914 to meet the Bloomfield's representatives. There are no surviving minutes of that meeting but it is fair to assume that the discussions were rather tense. Because of the poor financial state of the syndicate, Bloomfield's decided to forgo all the fees due to them for that first year but informed St Petersburg that they would take sole ownership of the steam drifter *Pesha*. Maybe the Russians were unaware that James had already taken this action some five months previously!

HERRING PEOPLE: 'Wee' Green

William James Empson Green was born in Winterton, Norfolk, in 1877 into a fishing family and first went to sea when he was 10 years old. For much of his adult life William weighed between 18 and 20 stone – hence his nickname, 'Wee'.

Wee became a steam drifter skipper, probably working for the Smiths Dock Trust Company, where he met James Bloomfield. The two men obviously hit it off because in 1907 they went into a 50/50 partnership (a very profitable one, as we have already seen) in a new steam drifter, the *Ocean Gift*. When James founded Bloomfield's Limited they sold the *Ocean Gift* and 'Wee' joined the new company as a skipper and half owner of the brand new drifter *Ocean Reaper*.

When James needed an English skipper for the *Pesha*, Wee Green was his first choice. The fact that Wee accepted this challenge shows considerable courage and enterprise, as he had probably never previously sailed outside British waters. In later years he rarely spoke of the Russian trip to his family and the only tangible mementos of the trip are his specially-issued passport and his contract of employment for the trip.

Wee Green was from the Norfolk fishing village of Winterton. As well as the White Sea voyage on the *Pesha*, his other claim to fame is the development of anti-submarine nets during the First World War, when he served in the Royal Naval Reserve.

Timelines Chapter 3

1911 Russia imports 3.3 million cwts of British salted herrings (2012 value; £152 million)

1912 Bloomfield's Limited are approached by the Russians

1913 Bloomfield's directors travel to meeting in St Petersburg

1913 Bloomfield's buy the trawler *Melbourne* for £800 and rename her *Pesha*

1913 'Wee' Green skippers the *Pesha* to Archangel

1913 Russia imports 3.6 million cwts of British salted herrings (2012 value; £187 million)

1913 The *Pesha* returns from Russia and is renamed *Ocean Comrade*

1914 Bloomfield's claim ownership of the *Pesha*

CHAPTER 4

1914 to 1919 – War!

For a herring-selling nation, there was probably no worse scenario than the United Kingdom becoming embroiled in a war that involved the two main herring-buying countries. On 1 August 1914, Germany declared war on Russia and three days later Great Britain declared war on Germany. That same day the order was issued for all British fishing vessels to return immediately to their home ports.

For centuries, successive monarchs and governments had believed that, as an island nation, Great Britain needed men with maritime experience who could be called up for naval duties in the event of war. The most obvious source of men and vessels was the fishing industries, so it had always been regarded as important to promote a large, successful British fishing industry. With war declared, the Admiralty now had an urgent requirement for men and for large numbers of small vessels to be used in a variety of support and operational roles. With their shallow draught, their speed and manoeuvrability, steam-powered fishing boats were ideal. On 4 August, the Admiralty began the process of commandeering steam drifters and trawlers from their owners. Because of their greater size and engine power, steam trawlers were the first choice for the Navy but it was soon realised that using steam drifters for support work would release trawlers to perform more active roles.

Some fishermen joined the Royal Navy but many were too old to serve in the regular armed forces and they were encouraged to enlist as ratings in the Royal Naval Reserve (RNR) and to serve on the commandeered fishing vessels. For many fishermen, especially those in the north of Scotland, once the fishing boats had gone, there was little alternative employment so many of them signed up for the RNR. By 1917 some 2,500 British fishing skippers were employed as RNR Skippers, 100,000 fishermen were serving in the RNR and three quarters of all first class (over 25 tons) fishing vessels were in Admiralty service, including 1,400 steam drifters.

Bloomfield's Goes to War

The Admiralty quickly commandeered Bloomfield's entire fleet of steam drifters and trawlers and many of Bloomfield's crews enlisted in the Royal Naval Reserve as Skippers or ratings (James Bloomfield is reputed to have personally driven his car to their homes so he could take them to the recruiting office).

As Bloomfield's Yarmouth headquarters was no longer needed (they had no fleet left to manage), the company allowed Lady Savile-Crossley from nearby Somerleyton Hall to use the buildings as a war hospital. On 19 January 1915, Yarmouth became the first British town to sustain serious damage and loss of life from the air when Zeppelin L3 flew over the town, dropping bombs. Considerable damage was caused on South Denes Road, where the Fish Warf sustained damage valued at £550 and the Refreshment Rooms, £125. After seeing the destruction in the town following the bombing (two people were killed), the authorities decided to move the hospital inland and away from military targets.

HERRING PEOPLE: Wee Green

At the outset of the War, Wee Green, like many other drifter skippers, joined the RNR. As German submarines increasingly showed how effective they were in destroying shipping, it became important to devise ways of counteracting the menace that they posed. Drawing on his experience with driftnets, Wee proposed the design and development of anti-submarine nets. Like herring nets, these were suspended curtain-style below the surface but were made of steel. They were effective in protecting shipping in harbours and at anchorages such as Scapa Flow. Because he was a serving member of the forces he did not receive any reward or recognition for this important contribution to the war effort. By the end of the War, Wee had attained the rank of Chief Skipper and had a number of vessels under his command.

HERRING PEOPLE: James Bloomfield's War

With no fishing fleet to be managed but being over-age for active service, James decided to lend his engineering experience to the war effort. The board of directors agreed to this and also that James should continue to receive a proportion (80 per cent) of his £1,000 annual salary from the company. He worked for the Ministry of Munitions in Coventry, Leeds and South Wales, for which he received a salary of £400 a year. His work was so highly regarded that the Admiralty recruited him for work in connection with minesweepers (i.e. steam drifters!). James suffered a family bereavement in 1915 when his brother Thomas was one of the 1,200 people lost when the *Lusitania* was torpedoed off Ireland.

At the end of the war, James was awarded the naval rank of 'Honorary Commander' in recognition of his efforts. He must have enjoyed this title because he continued to use it after the war, although the word 'Honorary' somehow became mislaid.

The Drifters' Naval Duties

In the first days of the war, the German navy laid hundreds of mines in the North Sea to prevent shipping using the ports of the Thames, East Anglia, the Humber and the Tyne. Some of these mines were anchored to the sea bed and others floated, carried by the tides. During the course of the war the Germans would lay some 25,000 mines around the United Kingdom coast and they were a hazard not just to warships and cargo vessels but also to small fishing boats sailing to and working on the fishing grounds. Many of the drifters now in the RNR were used as mine-sweepers; this meant working in pairs with a net between them to 'catch' mines and to then detonate them, usually by shooting at them with rifles. This was dangerous work. The first steam drifter to be lost in the war was probably the *Eyrie* of Lowestoft, blown up on 2 September 1914 while engaged on minesweeping duties off the East Anglian coast.

Some drifters were deployed to set and patrol anti-submarine nets while others were employed on boom defence work or acted as tenders in ports and

During the First World War, James Bloomfield worked as a civilian for the Ministry of Munitions and then for the Admiralty. For his services to the Admiralty he was given the rank and uniform of an Honorary Commander.

fleet anchorages, ferrying personnel between ship and shore and delivering stores, water and fuel. Most were fitted with a naval gun and the crews were trained to use them. Generally speaking, the vessels on minesweeping duties had a 6-pound gun and those on other duties had a 3-pounder. The steam drifter proved to be such a useful vessel that during the war the Admiralty had a further 125 built to a standard design.

War and Herrings

For those still fishing, life was difficult but the rewards could be good. Any vessels trying to fish in the North Sea ran the risk of either hitting mines or of catching mines in their nets. Because of this danger, parts of the North Sea were closed to fishing. A further risk was that German U-boats had been ordered to destroy any British fishing vessels that they came across in order to cut food supplies to the United Kingdom and reduce morale. In the early days, the U-boat commanders were gentlemanly about this, hailing the skipper and giving him and his crew the chance to get into their lifeboat before they sank the vessel. Later in the war they tended to just open fire. Several hundred British fishermen were captured by the German navy and spent the war in harsh conditions at a prison camp near Berlin.

To the crews still fishing, the Admiralty now offered a series of rewards for providing information on the movements of enemy vessels. The top reward on offer was £1,000 'for information which directly leads to the actual capture or destruction of an enemy vessel down to and including a minelayer or submarine'.

Fishing did continue but it tended to be mainly on the west coast, which was a bit safer. Some boats were lost to mines but herring continued to be landed, although the total catch was a fraction of peace-time landings. In Scotland the 1915 herring catch was just 15 per cent of that of pre-war days but in the following three years it rose to about 50 per cent. Not surprisingly, with only limited fishing in the North Sea, English herring landings had plummeted to less than 10 per cent of those in 1913. However, the value of this catch was 30 per cent of the 1913 figure, indicating that wartime shortages had tripled the market prices. In the relatively safe waters of Loch Fyne, Scotland, some fishing boats were reported to have earned £3,000 in a year and some of the fishermen £750 each. Needless to say, during the war exports were practically non-existent.

The fresh fish trade experienced difficulties during the war because the railways had to give priority to military traffic. This meant that fish trains often took a long time to reach the cities, resulting in a large increase in the quantities of fish at Billingsgate being condemned as unfit for consumption.

The wartime shortage of herrings led to the appearance of the 'Painted Lady', an artificially coloured kipper. Herring smokers were usually able to produce 16 stones (224 pounds) of kippers from a cran of herrings (herrings were sold by volume but kippers were sold by weight). During the war some kipper producers decided to cut down on the smoking time because the longer the kipper was in the smoke-house the more moisture (i.e. weight) it lost. The shorter time in the smokehouse produced a very light-coloured kipper (it is the exposure to smoke that makes a kipper brown) but to ensure there was no loss in colour and so that the customer would not notice the difference, the smokers first coated the kippers in annatto, a red vegetable dye. The producers who did this found that they could now get 18 stone (252 pounds) of kippers from a cran of herring (an extra 12½ per cent). As more kipper producers jumped on the band wagon, smokers who wanted to produce traditional dye-free kippers were priced out of business and for the next 25 years the consumer became accustomed to the very red kippers which were available in the shops.

The absence of British herrings from the international market was a boon to Holland and Norway because both these countries remained neutral throughout the war and were able to carry on fishing. The Germans could no longer buy British herrings so they bought a large part of the Dutch catch. In 1915, with food imports restricted by the British naval blockade, Germany bought a fleet of second-hand trawlers from Norway with a view to becoming more self-sufficient in fresh fish.

Germany also bought large quantities of both herring and cod from Norway and in 1916 British diplomatic staff were instructed to disrupt this trade. They were assisted in this task by John Irvin of Aberdeen (son of Richard Irvin – see Introduction), who was one of a team of advisers to the Board for Agriculture and Fisheries. Working from Stockholm, the British managed to buy up three quarters of the total Norwegian salted herring production simply to prevent it going to Germany. They spent £11 million on these herrings and then sold them to Russia for half of that sum. Unfortunately for John Irvin, the ship on which he travelled back to the United Kingdom was arrested off Rotterdam by the German navy and he was interned (along with large numbers of captured British fishermen) in Ruhleben prison camp outside Berlin. The following year he was awarded a knighthood in recognition of his efforts.

Drifter Heroes

For those British drifter skippers who were now on naval duties, life could be exceedingly dangerous. It was one thing to be fetching and carrying in a sheltered anchorage such as the Solent or Belfast Lough but it was a very different matter if you were escorting convoys of larger ships on the open seas. Many steam drifters were assigned to the Dover Net Drifter Flotilla, better known as the 'Dover Patrol', which was based in Dover and Dunkirk. The purpose of this force was to escort and protect British shipping, to maintain the supply route from the Kent ports to northern France and, with mines and anti-submarine nets, to keep the

English Channel closed to the German Navy. The Dover Patrol was attacked so often that it gained the dubious nickname of the 'Suicide Club'. In just two days in October 1916 six steam drifters were among vessels lost in the Battle of Dover Straits.

Elsewhere, two drifter skippers were awarded the Victoria Cross (VC) for their gallantry during the war. One was a posthumous award to Thomas Crisp of Lowestoft. He was a RNR Skipper in charge of the armed sailing smack *Nelson*. The role of this vessel, a Special Service 'Q-ship', was to be a decoy in the North Sea; she looked like any other fishing boat but was fitted with a 3-pound gun so she could attack German submarines that surfaced near her. In August 1917, the *Nelson* became involved in a very one-sided exchange of fire with a surfaced U-boat and although grievously injured, Tom Crisp refused to surrender and went down with his vessel. He had previously been awarded the Distinguished Service Cross for an earlier clash in which he had succeeded in sinking a U-boat.

The second VC was won in a naval action in the Straits of Otranto, the narrow stretch of sea where the Adriatic meets the Mediterranean. Steam drifters were maintaining an anti-submarine barrage in May 1917 when they were attacked by three light cruisers of the Austro-Hungarian navy. Although under heavy fire, the drifters stood their ground and continued to fire their guns at the larger enemy ships. Fourteen drifters and sixty-seven crewmen were lost in this action. Joseph Watt of Fraserburgh, despite his drifter the *Gowanlea* sustaining heavy damage, continued to attack the cruiser *Novara* (crew of 318) and was awarded the VC. Two other skippers involved in that action each won the Distinguished Service Cross. After the war, the *Gowanlea* was sold to Hopeman owners.

HERRING PEOPLE: Joseph Watt

At the outbreak of the Great War, Joseph Watt was the skipper and owner of the Fraserburgh steam drifter *Annie*. He joined the RNR, where he became skipper of the drifter *Gowanlea* and sailed her to the Mediterranean. In 1916 he took part in the evacuation of the remnants of the Serbian army from Albania and for this he was awarded the Serbian Gold Medal for Good Service. As well as the VC, he also received the *Croix de Guerre* and the Italian Silver Medal for Military Valour.

A quiet, reserved man, he returned to herring fishing after the war and he rarely mentioned his medals – indeed, they were kept in a drawer of bits and pieces on board his drifter. At the outbreak of the Second World War he again volunteered and once more skippered a naval drifter, but this time served only in British waters. He died in 1955 at the age of 67. In 2012, his collection of medals from the two wars, including his Victoria Cross, was sold at a London auction where they fetched £170,000 (plus buyer's premium of 20 per cent).

Money Matters

Owners of requisitioned boats had lost their main source of income but by way of recompense, they received a charter fee from the Admiralty. Once it was obvious that fishing would be severely restricted, some boat-owners even became keen for the Admiralty to take their vessels and so ensure them a modicum of income. Initially the charter fee for steam drifters averaged about £475 per year, depending on size and condition, but owners felt this was too low. Bloomfield's Limited received varying fees from the Admiralty for their fleet. For their newly built drifter *Ocean Gain*, they were paid £530 per annum (considerably less than the £3,300 gross income which, as we saw in Chapter 2, Bloomfield's had been earning from her sister vessel *Ocean Angler*).

On behalf of the industry, James Bloomfield and his fellow director, Charles Goldman, lobbied for an increase in charter fees. They were both ideally placed to do this as Goldman was still a Member of Parliament and James was now working for the Admiralty. In response to this pressure, in 1916 the Admiralty introduced a new formula to calculate charter fees. This was based on a value per ton combined with the horsepower of the vessel and, as a result, the annual charter fee for drifters rose to an average of £670 per vessel.

Insurance was a big worry for the owners while their boats were under Admiralty charter because war risks were excluded from normal insurance cover. To placate the owners, the Board of Trade founded the British Fishing Vessels War Risks Association Limited to underwrite the insurance industry for any losses incurred while fishing boats were on naval duty.

The driftermen now serving as RNR ratings earned between 7 shillings and 4 pence (37p) a day (Engineers) and 4 shillings and 10 pence (24p) a day ('Other Crew'). Pay rates in distant areas like Orkney and Shetland were 15 per cent higher. These levels of pay were not a lot for men who until 1914 had been making a good living from herring fishing.

Bloomfield's Russian Venture

Initially, the war had little effect on the Russian Venture but Bloomfield's were probably relieved that they at least had the *Ocean Comrade*, although this vessel had been commandeered with the rest of the Bloomfield fleet, converted to a minesweeper and allocated to the Dover Patrol.

By 1916 the Great War had been in progress for two years and things were not going well for Russia. In Yarmouth the confidence of 1913 was rapidly evaporating. Bloomfield's had still not received any dividends from Russia, so they decided it was time to pull out of the venture. They appointed a firm of accountants in Petrograd to dissolve the partnership. (At the start of the war the Russians felt that 'St Petersburg' sounded too German so they renamed the city 'Petrograd') The accountants managed to sell the Bloomfield's holding to the syndicate's Russian partners and after deducting the value of the *Pecha/Ocean Comrade*, they obtained 163,000 roubles. This sum seems reasonable, given that Bloomfield's original investment had been 110,000 roubles. The money was placed in Bloomfield's bank account in Petrograd to wait until the war was over and for normal international bank transactions to be resumed.

In 1917, Petrograd witnessed the beginning of the Revolution and Russia then descended into a bitter civil war. In the wake of the Revolution, international trade with Russia collapsed and Bloomfield's money just sat in the Petrograd bank account, and it appeared each year on the Bloomfield's balance sheet of as an asset to the value of £11,770 (the sterling value of their original investment).

HERRING PEOPLE: Tom Bruce's War

In the first twelve months of the war, most of the Hopeman steam drifters were requisitioned by the Admiralty and their crews enlisted in the Royal Naval Reserve. The brand new *Admiration* and her skipper, Tom Bruce, were sent to Scapa Flow, Orkney, where for the duration of the war she was used as a boom defence vessel and a water carrier. Tom's RNR pay was probably enough to live on but he and his partners struggled to pay back their loans out of the charter money that they received from the Admiralty.

Tom's father, John, also joined the RNR (at the age of 50) when the steam drifter *Charity* was requisitioned for use as an anti-submarine net vessel. On Sunday 24 October 1915, with a crew of seven ratings on board, John sailed the *Charity* out of Great Yarmouth harbour to be fitted out in Poole, Dorset. The boat was never seen again, 'lost, presumably mined'. No bodies were ever recovered.

Return from War

Throughout early 1919 fishermen were released from RNR service and the Admiralty began to return drifters and trawlers to their owners, who resumed fishing with them. However, 675 fishing boats, including 130 steam drifters, had been lost. As the Scotch Voyage got under way in June 1919, there was an expectation that the industry would soon return to those hectic pre-war days. The government (prompted by James Bloomfield and others) provided price subsidies for the 1919 English season to help get the fishermen back on their feet financially.

Bloomfield's Wartime Financial Performance

Although Bloomfield's Limited had carried out no fishing during the war, it had continued to function at an administrative level – this had included some fish-selling activities. Most of the directors were busy on war duties but they still received their directors' fees, their dividends and, in the case of James Bloomfield, most of his salary. Board meetings and annual general meetings were held on an *ad hoc* basis to sort out pressing problems such as the Russian venture.

Year	Directors' Pay	Net Profit	Dividend
1914	£1,836	£1,750	None
1915	Not recorded	£9,955	7½%
1916	£2,773	£17,581	10%
1917	£1,906	£11,718	10%
1918	£419	£9,000	10%

Table 10 Bloomfield's wartime financial results.

As can be seen from these figures, the investors in Bloomfield's by and large had little to complain about during the war.

HERRING PEOPLE: William Lever

For most of his life, William Hesketh Lever had no connection at all with the herring industry. He was a typical Victorian self-made industrialist. He was born in 1851 in Bolton, Lancashire, son of a wholesale grocer. He left school at 15 to join his father's business, where he increased turnover and became a partner. Not one to rest on his laurels, in 1884 William registered the trademark 'Sunlight Soap' and formed the soap manufacturing company Lever Brothers Ltd.

He had a very strong work ethic and a forceful personality (he seldom took 'No' for an answer). He was a non-drinker, a non-smoker and never gambled. Time was so precious to him that he had his household plumbing specially adapted to fill his bath with hot water in seconds so he did not have to waste the five minutes it normally took to fill. He became Member of Parliament for the Wirral in 1906 and was created a baronet in 1911, becoming Lord Leverhulme.

By 1916, Leverhulme, who was in his late sixties, was struggling to maintain his tight personal control over Lever Bros. He was looking for a new project which he could run without interference from board members and banks. In 1917, while on a sentimental visit to the Hebrides (he had honeymooned there many years previously), he saw that the Isle of Lewis was for sale and so he bought it. Seeing the poverty on the island, he began looking at ways that he might develop its economy for the benefit of the island's servicemen when they returned from the First World War. Realising that the island's only natural resource was the plentiful fish stocks, particularly herring, in the surrounding seas, he hatched the idea of developing an industrial-style fishery to create jobs for the local people. He quickly gained an understanding of the workings of the fishing industry and started thinking about how he might apply modern business thinking and industrial methods to it.

He expanded the facilities at Stornoway harbour by building a canning factory and a fish reduction plant (to produce fertiliser). Among his more progressive ideas was a proposal to use aeroplanes to spot the shoals of herring and to guide the fishing fleet towards them. He also believed there was a big future for frozen fish but despite conducting trials at Fleetwood, he was unable to convince the general public.

He recognised that the salted herring trade with northern Europe was not going to quickly recover to its pre-war levels but believed that home sales of fresh herring could be increased. He realised that the best way of ensuring there was a ready market for the fish you have caught was to have your own chain of retail fish shops. The remoteness of Lewis was a problem but he planned to overcome that by having a fleet of refrigerated steamers carrying the fish to Fleetwood. From there they could be easily transported by rail around the industrial north-west.

1919 – Lord Leverhulme Moves into the Fishing Industry

During the early part of 1919, Leverhulme set about using his considerable personal fortune to found a private company which he eventually named Mac Fisheries Ltd. This company would combine fishing, fish processing and retailing and would be entirely divorced from Lever Brothers. Rather than start from scratch, he bought up a number of fish-related businesses that had proven track records. Among these were a Billingsgate wholesaler, several herring curing firms (including that of Andrew Bremner at Wick) and, from Richard Irvin & Co., the Southern Whaling & Sealing Co. He also bought 28 acres of land from the Borough Council in Yarmouth to be used as herring curing yards. As this land included part of the town's race course, the Council had to relocate that popular amenity to a new site on the north end of the town.

Leverhulme also needed a fishing fleet and, wanting the best, he set his sights on Bloomfield's Limited. On 3 September 1919, James Bloomfield informed his directors that Lord Leverhulme had offered £3 per share. By 1 October, James had managed to push this up to £3 2s 6d before advising shareholders to accept Leverhulme's offer. The company became a division of Mac Fisheries but retained the name 'Bloomfield's' and James remained as managing director and chairman of the board.

Meanwhile on Lewis, Leverhulme had misjudged public opinion. When the island men came home from the war, what they wanted was land for crofting (the land that Leverhulme now owned) and not his fishing jobs. Leverhulme quit Lewis and bought neighbouring Harris. Here, with the support of local people, he began to construct a large port and fish processing facility at the tiny village of Obbe, which in 1920 was renamed Leverburgh. By 1921 he had completely severed his connections with Lewis but continued to plough his energies into Mac Fisheries and into building up Leverburgh.

Russia – The End of the Story

At the annual general meeting of Bloomfield's Limited on 21 February 1919, James Bloomfield reported that the *Ocean Comrade* (formerly the *Pesha*), which had recently been released from Admiralty duties, had been sold to a new Scottish owner for £4,500. This was a good price for a twenty-seven year old vessel which by then had a 'book value' of £800 and shows that James could be a very astute businessman.

At the same meeting, he reassured the shareholders about a balance sheet item, the rouble asset of £11,770, held in Russia. He said '… the last information that we have had from Deloitte, Plender and Griffiths was that they had the highest opinion of the National City Bank of New York in Petrograd.'

Six months later, the rouble slid into a period of hyper-inflation. At the start of 1920, only weeks after the Leverhulme takeover, Bloomfield's auditors insisted that the 'roubles asset' be written off as it 'now had a market price of nil'. We can only imagine Lord Leverhulme's thoughts on this turn of events, especially as Lever Bros had recently paid £8 million for the Niger Company of West Africa only to discover that this new acquisition owed the banks £2 million!

Timelines Chapter 4

1914 Most British steam drifters are requisitioned by the Admiralty

1914 James Bloomfield joins the Ministry of Munitions

1915 The drifter *Charity* is mined with John Bruce and his crew of seven aboard. No survivors

1915 Germany buys a fleet of trawlers from Norway

1916 Britain buys most of Norway's herring catch

1916 Joseph Watt, skipper of the *Gowanlea*, wins a VC for action in the Straits of Otranto

1916 The Admiralty increases the charter fee for its use of steam drifters

1917 Tom Crisp, skipper of the Q-ship *Nelson*, is awarded a posthumous VC
1917 The Russian Revolution begins in Petrograd (formerly St Petersburg)
1917 First use of the vegetable dye annatto on British kippers
1919 Lord Leverhulme buys up fishing businesses to form Mac Fisheries
1919 Lord Leverhulme buys 28 acres in Yarmouth for Mac Fisheries
1919 Bloomfield's sell *Ocean Comrade* for £4,500

CHAPTER 5

The 1920s

The Boom Years are Over

In March 1919 James Bloomfield warned his shareholders that reconstruction of the herring industry was not going to be easy because before the war '97% of the trade had gone for export'. After the upheavals of the war, both Germany and Russia had undergone major changes economically and politically. As Germany struggled to pay reparations and to rebuild its economy, it ran into a notorious period of hyper-inflation which played havoc with international currency dealings. Herring curers who in 1922/3 had sold herrings to Germany and had been paid in German marks were caught out by the rapid inflation of that currency – one Wick curer is reputed to have papered his sitting room walls with the now worthless notes that he had received for his herrings.

Russia was a rather different case in that one of the aims of its new Communist leaders was to destroy world capitalism and, therefore, they were reluctant to engage in trade with the capitalist West. In 1919 Lenin had deliberately prompted the hyper-inflation of the rouble which hit Bloomfield's Limited so hard. The only herring imports that Lenin would now allow were poor quality salted herring from Norway, which Russia could buy for half the price of Scottish herring. Because of Russia's politics, there would be occasions during the 1920s when, for ideological reasons, the British government discouraged any trade with the Soviets.

At home there was a feeling that the country owed a lot to our gallant fishermen for their wartime sacrifices and, in order to help them, the government bowed to pressure and during 1919 and 1920 set a minimum quayside price for herring in East Anglia. This helped the fishermen but it left some curers holding large stocks of barrelled herrings which they were unable to sell. As the general economic downturn of the 1920s began to take hold, the government was no longer willing to continue to give special assistance to the herring industry.

Bloomfield's Get to Grips with the Downturn

In 1921, in common with the rest of the industry, Bloomfield's Limited began a rigorous cost-cutting programme to adjust to the new economic climate. That year, their bank overdraft stood at £335,000. All deeds and ownership certificates had to be lodged with the bank as security. Tangible assets such as boats, buildings and equipment were re-mortgaged to raise cash to help reduce the overdraft. In 1922 they had to close down a number of departments and branches that James Bloomfield had built up only a few years previously. Among those to go, with resulting loss of jobs, were the Stores Department, the Kippering Department, the Buckie branch office and the Yarmouth and Lowestoft trawl offices.

If a large company like Bloomfield's was struggling to remain solvent, one can guess how hard it must have been for smaller companies, sole traders and owner-skippers who did not enjoy the asset-base and reserves of a large company like Mac Fisheries. The pruning carried out by Bloomfield's seems to have worked because, despite the continuing depression in the herring industry, by 1927 Mac Fisheries were making an annual profit of £137,000. Mac Fisheries was absorbed into Lever Brothers in 1922 and remained part of that company, even after Lord Leverhulme's death in 1925.

While Mac Fisheries were under pressure to cut costs at Yarmouth, the company's owner, Lord Leverhulme, was pouring his own money (an estimated total of £500,000 – £30 million today) into Harris, in the Outer Hebrides. Here he was pursuing his personal aim of developing Leverburgh, where he wanted to build a harbour capable of accommodating fifty drifters and a fish processing centre. Leverburgh opened for business in 1924 and twelve Bloomfield's drifters made such huge landings that additional staff had to be quickly recruited from the mainland in order to process all the herring. When Leverhulme died in 1925, the Board of Lever Bros ordered all construction work at Leverburgh to cease, sacked the staff and sold the harbour site for £5,000.

HERRING PEOPLE: Vladimir de Sivers

In 1921 Bloomfield's received a letter from Vladimir de Sivers. In it he said that he had lost the certificates for the 2,000 Bloomfield's shares that he had bought in 1912 and still owned and asked for copies to be issued. The surviving records do not tell us where he was writing from but his letter confirms that at least one member of the Russian aristocracy survived the Revolution. There is no record of Bloomfield's reply – it would have made interesting reading.

James Bloomfield's former home in Yarmouth. This elegant property is close to the Yarmouth sea front and is now a hotel.

HERRING PEOPLE: James Bloomfield

James Bloomfield had occasionally suffered health problems; it is possible that the strain of his wartime duties followed by the stress of cost-cutting had further damaged his health. After May 1922 he became too ill to attend any more board meetings and on 4 November he died. The *Eastern Daily Press*, the *Yarmouth Mercury* and the *Fishing News* all printed glowing obituaries which highlighted his war service and business successes. They recounted how popular he had been with his skippers and how he had always campaigned strongly on behalf of the herring industry. There was no mention of Russia.

His funeral was attended by many of Yarmouth's most prominent figures including the Mayor, the Town Clerk and representatives from many fishing companies. His old friend Wee Green was one of the pall-bearers. James had become a wealthy man; in addition to his salary, dividends and earnings from other sources, the sale of his shares to Lord Leverhulme would have netted him the modern equivalent of over £2 million.

Scotland in the 1920s

The 1920s were a tough period for the entire herring industry but they were especially hard in the Moray Firth fishing towns and villages. Fishermen had borrowed heavily before the war to buy steam drifters which had been commandeered for five years and many still owed money on their vessels. It was now a struggle to make a living from these boats. They were proving costly to run, new nets were expensive and herring prices were low – it was all so very different from the boom years. Drifters that before the war had cost them £3,000 were now selling for as little as £150. This put the owners in a dilemma; if they sold the boat, the price they received would be less than the outstanding debt on it but without a boat they would still struggle to make a living because there was no alternative employment. Because share fishermen were ineligible for unemployment benefit, for many of the Scots the least bad option was to remain in herring fishing and to limp along with increasingly worn-out equipment in the hope that there would soon be an upturn in the herring market. Some fishermen saw migration as their best option and during the 1920s there was something of an exodus from the Moray Firth area to Canada.

One solution for those owners who still had savings and were not heavily mortgaged was to take a loss on selling their steam drifter and to buy a motor fishing boat. The Moray Firth fishermen had seen Danish fishermen in the North Sea using the new seine net to good effect in catching cod and haddock and they wanted to follow suit.

In the Outer Hebrides the residents of Lewis may have begun to regret snubbing Lord Leverhulme and his promised fish-related jobs. In December 1923 *The Times* reported that 20,000 inhabitants of Lewis and Harris were threatened with starvation. The islands had been hit by a succession of weather-related disasters; the potato crop had failed, the hay had been ruined (so no food for the cattle), the harvest was destroyed and there was a shortage of fuel.

A feature of the early Twenties in the fishing areas of north-east Scotland was the emergence of a popular religious revival. This may have been triggered partly by the enormous loss of life during the war and partly by the current financial hardships. The Scottish Baptists, the Brethren and the Salvation Army undertook an evangelical recruitment which extended to sending preachers to Yarmouth and Lowestoft each autumn to recruit the visiting Scots fishing workforce. On Sundays in Yarmouth, the Scots workers would pack a number of local church buildings which the Scottish churches took over for the season. There were often open-air prayer meetings with sermons and hymn singing which could last for up to four hours.

Herring Fishing Perseveres

In an effort to keep up supplies of herring for the home market, fishermen and scientists looked for new herring stocks and populations which could be exploited and so, for example, some drifters now took part in a spring herring fishery off East Anglia. Despite falling exports, the herring industry continued to employ many thousands of people and it still brought economic benefits to the large fishing ports. The table below shows the additional jobs during the herring season at Yarmouth Fish Wharf.

Role	Number, Jan–Sept	Number, Herring season
Labourers	3	12
Lavatory Attendants	1	3
Berthsmen	1	10
Basket Stewards	–	2

Table 11 Numbers employed at Yarmouth Fish Wharf.

As in better times, the volume of cargos sailings from Yarmouth and Lowestoft rose during the autumn as Klondykers and other cargo vessels took away herring. For most of the year, cargo sailings from Yarmouth tended to be to local destinations such as Hull, Newcastle and London but during the autumn, the list grew to include the ports of Danzig, Hamburg, Altona, Königsberg, Duisberg, Memel and Libau.

Statistics for the 1920s show that average herring landings per vessel at the Yarmouth season were around half of those in 1913. In addition, the price of coal and of new nets had risen but herring prices had dropped, so it was increasingly difficult to keep herring drifters in profit. It was reported in 1922 that 800 Yarmouth fishermen had had no work since 1920. In 1927 Horatio

A cargo boat loading up barrels of herring. All herring ports saw regular cargo sailings as vessels arrived to load barrels of herring for export. The *Hama* is pictured at Yarmouth in the 1950s.

Fenner Ltd, once one of Yarmouth's largest drifter-owning companies, was wound up. This pattern of lay-offs and closures was repeated around all the herring ports.

In 1928 an ominous landmark was reached when the total German herring landings (during the war they had bought a fleet of trawlers from Norway) equalled the total East Anglian catch. This was bad news for the British trade, who had already seen the suspension of the Russian trade, because it meant that Germany would now need to import fewer herrings.

HERRING PEOPLE: Andrew Bremner

After Lord Leverhulme bought him out in 1919, Andrew Bremner had led a much quieter life. He spent a lot of time on his estate at Keiss but he could often be seen in Wick, taking an interest in former colleagues and in the herring industry. He was still active but was no longer in the best of health and on 15 October 1923, he died aged 71. His estate was valued for probate at £101,000 (over £19 million at today's values) and he was probably one of the last of the old-style Scots curers to die with his fortune intact.

HERRING PEOPLE: Tom Bruce

During the 1920s Tom Bruce was one of many Scots fishermen who struggled to make a living from herring fishing. He stayed with it because he could not afford to leave (he still owed money he had borrowed in order to build the *Admiration*). By 1924 he and Jessie had four children and his widowed mother to fend for. Home was now the top two rooms of a small house overlooking the Hopeman railway sidings. Their two sons had to sleep in a relative's house across the road.

Tom's youngest daughter, Chrissie, remembers there being very little money in the house during the 1920s. A staple food was salted herrings from the barrel that stood in the shed; boiled herring with boiled potatoes made up the Saturday meal. In between seasons, the bedroom Chrissie shared with her sister, Ella, became a net repair workshop where their mother and granny spent many hours trying to repair nets that ought to have been replaced by new ones. At these times the two girls slept surrounded by nets and the debris (bits of dried jellyfish and so on) that was still attached to them. Chrissie recalled the deep gloom around the house one year when Tom returned early from the Shetlands, having lost most of his nets in bad weather. Happier memories were of the present that he brought back from Yarmouth each year – a box of apples.

Tom's brother-in-law, Alec, formerly skipper of the drifter *Devotion*, emigrated from Hopeman to Canada but soon returned after finding things were just as tough out there.

The Continuing Dangers of Herring Fishing

By the 1920s the number of casualties at sea had fallen from the levels of the nineteenth century but the records of the Great Yarmouth Fishermen's Widows and Orphans Fund show that lives were still being lost. By now the payments from the Fund had risen to 10 shillings a week for widows and 5 shillings for a dependent child. The Fund was an early victim of the tougher economic times in the industry when in the early 1920s its directors reduced the levy on boat owners (which represented the Fund's only income) from £3 a year to £2 a year for each boat.

Apart from loss of life, there were a host of other mishaps at sea and the records of the Lowestoft Mutual Fishing Vessels Insurance Society Ltd (Appendix 4) show how many lucky escapes there were during the 1920s. While many of these incidents may not have been life-threatening, they are a reminder of the uncertainties of the fishing industry and also of how far away from home these vessels were working. If a steam drifter lost power at sea it would need to be towed to port, often by another fishing boat. In such cases the rescuing vessel was entitled to make a claim for salvage money, the amount of which was usually disputed by the insurer. In the case of the *Lord Fisher*, their rescuer, another fishing boat, claimed £500, was offered £75 and eventually settled for £175.

It was not unusual for other shipping to run into drifters that were busy fishing. In 1927 the Lowestoft drifter *John Alfred* was riding to her nets off the Durham coast when she was hit and badly damaged by a mystery ship which failed to stop. Investigations suggested that it might have been an Estonian collier, the *Pakri*. The *John Alfred*'s insurers had the *Pakri* arrested while she was in port in London and the crew were interviewed. After a court appearance the following year, a financial settlement was reached.

Bloomfield's International Sales Push

After the death of James Bloomfield, Neil Mackay was appointed the new General Manager. He had considerable experience in the herring trade, having previously been manager of the Scottish herring exporting company Duncan & Jamieson, which had been another of Lord Leverhulme's 1919 acquisitions. In the months that followed the war, the map of Europe had been re-drawn and several new states were created from the old Germany and Russia. The new herring-consuming states which now had their own identities included Poland, Latvia, Lithuania, Estonia, Finland and Czechoslovakia. New herring trading networks had to be established to supply them.

Bloomfield's were now responsible for marketing the salted herrings produced by curing companies Andrew Bremner Ltd, James More Ltd, Joseph Slater Ltd and Wm Low Ltd (all now part of Mac Fisheries). Selling herrings had been relatively easy pre-war but it now became really tough. Not only had the Russian trade collapsed but during the hostilities the Norwegians and Dutch had grabbed a large share of the trade and the Germans had built up their own herring fleet. Bloomfield's had their own network of foreign contacts and soon opened new sales offices in Stettin and Danzig in order to supply the Polish market. In 1921, they opened a new branch in Berlin but the following year they had to write off a loss of £1,000 incurred there through falling exchange rates on the German mark.

In 1924, Bloomfield's opened a sales office in New York, where there was a steady demand for salted herrings among people who had emigrated from the Baltic States. Bloomfield's registered 'Revolver', 'Pistol' and 'Mallet' as brand names for their cured herrings in the United States.

In 1928, in an attempt to find new markets for their herring, Bloomfield's sent James Johnston, their man in Danzig, to Romania to explore the possibilities for selling cured herring. Johnston reported that it was a very poor country but there was a demand for Scotch and Yarmouth Matfulls and Mattie Firsts. There was, however, strong competition because the Norwegians had already established trading links there. The main difficulty in selling herring to Romania was the high rail freight charges as all herrings would have to be transported overland from Danzig. Mr Johnson reported: 'I had a look at the Fish Market in Bucharest, an evil-smelling place, where all sorts of weird fish were on sale, most of them dried. I saw nothing to compare with the herring.'

Coaling the Drifters (Yarmouth in the Late 1920s)

One beneficiary of the pre-war rush for steam drifters had been the coal trade. Coal merchants now followed the herring fleet around the British Isles. In Scotland during the summer, they sold coal from hulks moored offshore which were replenished by colliers from Northumberland and Durham. At the East Anglian season, the coal companies employed Scotsmen as 'runners' (sales representatives) who would board the Scots drifters as they docked and negotiate price to get an order for coal. If successful, they received a commission on the sale.

The coal was quickly delivered to the vessel at a convenient point along the quayside. Each coal company employed teams of four 'layers off' (also known as 'coalies'), casual staff whose job was to deliver the coal from the merchant's lorry on to the boats. Each man had to carry on his back sacks containing 10 stone (63.5 kg) of coal up a narrow gangplank to shoot it down into the bunker. An average delivery of 5 tons would comprise eighty of these sacks. The coalies had to be both strong and nimble-footed and they worked from seven in the morning until very late at night. Saturday was their hardest night because the Scots boats had to be replenished by midnight, before the Sabbath began.

Other labourers were employed in the coal sidings to fill sacks from the railway wagons and to load them on the lorry. The goods yard at Yarmouth's Beach

TELEPHONE 337. **TELEGRAMS: "Hansell," Gt. Yarmouth.**

G. H. HANSELL & CO.'S
COALS

FOR CHEAPNESS, ECONOMY, AND QUALITY.
SPECIALLY SUITABLE for STEAM DRIFTERS.

CARTED TO ANY PART OF HARBOUR, OR
EX STORES AT GORLESTON.

ALL ORDERS RECEIVE IMMEDIATE ATTENTION.

A supply of coal was vital for steam-powered vessels. Skippers and drifter owners needed to know where they could obtain fuel.

Station was the main arrival point for coal trains and at any one time during the season there could be 300 full rail wagons waiting there. Up to thirty of these wagons were moved each day along the quayside tramway to sidings behind the Fish Wharf.

The 800–900 steam drifters working out of Yarmouth required a total of about 7,000 tons of coal each week, all of which was shifted by hand. Behind the labourers was a team of clerks who recorded all deliveries and prices and ordered wagons of coal from the railway companies. Each Monday, invoices for the previous week's deliveries were presented to boat owners or their agents. 1921 saw the first of many coal strikes during the Twenties which caused coal to be rationed. This hit the coal merchants and the herring industry hard. That year drifters were limited to 50 per cent of their normal supply and this was soon reduced to 25 per cent, making life very difficult for the fishermen.

The Disadvantages of Steam Power

In the boom years of 1900 to 1913 the steam drifter had been seen as cutting edge, 'must-have' technology but now the drawbacks of steam power were becoming apparent:

Coal was expensive (27–30 shillings a ton) and a drifter could use up to ten tons a week.
The coal bunkers took up valuable space on board the boat
The coal industry was becoming increasingly subject to strikes
Each drifter needed two dedicated crewmen to run the boiler and the engine

A sight common to all herring ports was the back-breaking work of coaling steam drifters. In the foreground two fishermen tip herrings from a cran basket into a swill (a basket used only at Yarmouth).

The steam boiler needed a weekly shutdown for maintenance, after which it took at least two hours to raise up steam again

So, although steam drifters still had many fans, fishermen now recognised that motor powered boats were more efficient and economical. Furthermore, the supply of fuel oil was more reliable than coal. By the late 1920s, orders for new steam drifters were already beginning to tail off and more motor-powered boats were being ordered.

HERRING PEOPLE: Wee Green

After the war, Wee Green built up his own small fleet of drifters while continuing to work as ship's husband to the Bloomfield's fleet. His W. J. E. Green Ltd boats were managed by Bloomfield's Limited until 1925, when Wee severed his links with that company. He had come to believe that you could not successfully run a fishing company from Mac Fisheries' head office in the centre of London.

In the 1920s he joined the committee of the Great Yarmouth Boat Owners Association, and of the Yarmouth Fishermen's Widows and Orphans Fund. He ploughed the earnings from his five drifter/trawlers into building five pairs of semi-detached villas in Winterton.

Technological Advances, 1920s

The scientific research and international sharing of information which had taken place at the beginning of the century at the Stockholm and Christiania conferences had been curtailed by the war. Research in the United Kingdom resumed after the war, with work carried out on herring populations by the laboratories at Lowestoft and Aberdeen, who used their research vessels and also gathered information from fishing boat skippers. One discovery was that it was possible to determine the age of a herring by counting the annual rings on its scales. This led to work on establishing the age mix of the fish in herring shoals, which assisted in the forecasting of future herring populations.

One of the Lowestoft scientists, Professor Hardy, developed a means of helping boats find herring. He knew that if there was no plankton around, there would be no herring, so if a skipper could detect plankton in a particular area he was more likely to catch herring there. He developed the Hardy Plankton Indicator, a metal tube which was towed behind the vessel, filtering the water. If the filter turned red or pink, it showed the presence of the plankton favoured by the herring. The Indicator was effective but very few skippers persevered with it, preferring to use their own skills and intuition.

In the 1920s the larger fishing companies decided that when updating their fleets, they would build dual purpose vessels, introducing the era of the drifter/trawler. This gave them the flexibility to quickly switch between the different types of fishing so that at the end of the Home Voyage, they could send their vessels trawling for a few weeks. Increasingly, motor engines (diesel and petrol) were being used in the fishing fleet. Radio technology had also developed and by the end of the 1920s some fishing vessels were fitted with wireless receivers which enabled skippers to listen to weather forecasts.

In the early 1920s the local herring fishermen on the Firth of Clyde and Loch Fyne refined the use of ring-nets. Ring-netting was an efficient way of herring catching that involved surrounding a shoal with a long net, a hybrid of a drift net and a surface trawl net, which required two vessels to work as a pair. The fishermen soon began to take this form of fishing to other areas such as the Hebrides. The Clyde fishermen were also responsible for introducing the 'feeler wire' in 1929. This was nothing more than a length of piano wire with a 5-pound weight on the end which was dangled over the side of the boat; by using the wire, a skilled deckhand could detect the presence of shoals of herring below the surface from the vibrations along the wire as they brushed against it.

The national media discovered during the 1920s just how photogenic the herring industry was and various short cinema newsreels were made by companies such as Pathé News that portrayed herring gutters and driftermen at work. With the increased use of photographs in newspapers, the public awareness of the industry grew. In 1929, John Grierson made his silent feature film *Drifters*, which vividly portrayed the hard working lives of herring fishermen. This film is acknowledged nowadays as the first full-length documentary ever made.

The Armistice Gale, 1929

The 1920s had been pretty harsh on the Scottish herring fleet but the final year of the decade was a real stinker. Having observed the Sabbath in Yarmouth and Lowestoft, the Scottish boats had sailed out at midnight on Sunday 10 November 1929, and were at the fishing grounds by Monday afternoon (11 November). The English fleet, who had been fishing during the Sunday night, came into port on Monday morning and while landing their herring they learned that a severe gale warning had just been issued. Sensibly, most of them decided to spend Monday night in harbour. The Scots, with no radios and, therefore, no warning, all had their nets out when the south-westerly gale suddenly hit. (It could take hours to haul a long train of nets back in)

Two vessels were sunk and two lives lost in the storm. In addition, many boats lost some or all of their fishing gear as ropes parted and nets were torn in the heavy seas and strong winds. Most of the lost nets were uninsured and many Scots had

The launch of a new steam drifter. In 1928 a new Bloomfield's steam drifter/trawler is launched at Selby. She was named *D'Arcy Cooper*, after the then Chairman of Lever Bros.

no spare set of nets. Forty boats returned immediately to Scotland as they did not have enough nets to carry on fishing and the remainder fished till the end of the season, but with a much reduced number of nets. This was yet another financial blow to a business sector that was already struggling. One estimate was that the total cost of replacing all the lost nets was between £100,000 and £150,000.

Timelines Chapter 5

1920 The Russian investment is written off in Bloomfield's annual accounts

1920 Two years of drastic cost-cutting begin at Bloomfield's Limited

1921 De Sivers asks Bloomfield's for copies of his share certificates

1922 James Bloomfield dies aged 54

1922 800 Yarmouth fishermen were unemployed in the early part of the year

1922 Mac Fisheries is absorbed into Lever Bros Ltd

1922 Bloomfield's Limited begin building dual purpose drifter/trawlers

1923 Hyperinflation in Germany causes the bankruptcy of some Scots curers

1923 Andrew Bremner dies, leaving an estate worth, at today's values, £19 million

1925 Robert Boothby is first elected as MP for Aberdeenshire and Kincardine Eastern

1925 Complaints in Yarmouth that the Scotch curers prefer to employ Scots – not local men

1925 Death of Lord Leverhulme

1927 The long -established Yarmouth fishing company Horatio Fenner Ltd is wound up

1928 The German herring catch exceeds the East Anglian catch

1929 The 'Armistice Gale' causes much damage to Scots boats off East Anglia

1929 Use of 'feeler wire' is introduced on some Scottish vessels to detect herring shoals

1929 John Grierson's film documentary *Drifters* is released in cinemas

1929 Lever Bros merge with the Dutch Margarine Union to form Unilever

CHAPTER 6

1930 to 1939 –
A Glimmer of Hope?

At the beginning of the 1930s the poor economic situation was worsened by the Wall Street Crash and the ensuing Depression, which made assistance for the herring industry even less of a priority to the British government. Between 1929 and 1933 the already low exports of British salt herring fell by 50 per cent. Sales to the newly independent countries in north Europe (Poland, Latvia etc.) had grown during the 1920s but now they too began falling.

As for Russia, trading relationships had now progressed from differences of ideology to difficulties over trade agreements and currency movements. The Russians were proving to be very tough negotiators and some years they would hardly buy any herring (preferring to buy from Norway or the Netherlands) and some years they would move in at the end of the season and buy at rock-bottom prices from curers who were selling at a loss. When you look at the annual herring exports to Russia (below) it is easy to forget that this was the country which before the war had regularly imported 3 million hundredweight a year of British herring.

Year	Cwts
1922	92,135
1923	13,545
1924	806,841
1925	195,284
1926	11,750
1927	213,509
1928	102,285
1929	33,000
1930	182,295
1931	37,616
1932	402,475
1933	5,283

Table 12 British herring exports to Russia in the 1920s and 1930s.

Despite the reduced trading (and earnings), the British herring industry soldiered on. Some people were still able to make a reasonable living but there were bankruptcies among both curers and fishermen. At each herring season around the country, familiar names still appeared, such as Andrew Bremner & Co., John Woodger & Sons, H. Sutton Ltd, Richard Irvin & Sons, J. Wolkoff and, of course, Bloomfield's, who were now operating twenty-two vessels. Scots boats still came to East Anglia each autumn even though their earnings were low and English boats still went on the Scotch Voyage. In 1913 fishermen were reputed to have been averaging £6 each per week but twenty years later they were lucky to be getting £1 a week. In Scotland, the traditional greeting to a drifter returning home after a herring season had now become 'Did ye clear?' meaning, 'Did you make enough to cover your expenses?'

English boat owners still had confidence in the industry and at the start of the 1930s some were still ordering new steam drifter/trawlers, unlike the Scots who had moved to building motor powered boats. In 1930 Harry Eastick of Yarmouth ordered a new steam drifter/trawler to be built from a local shipbuilder. They were very busy and contracted the job out to a boatbuilder in Kings Lynn and this vessel was the only steam drifter ever to be built on that side of Norfolk. Harry named her *Lydia Eva* after his daughter. In 1932, the launching in Aberdeen of *Wilson Line* marked the completion of the last steam drifter to be built in the United Kingdom. Everyone in the industry could now see that motor and diesel engines were the way ahead. Nevertheless, some of the steadily decreasing fleet of steam drifters would carry on working until the late 1950s.

The Curers Soldier On

Before 1914 curing had been the most profitable branch of the herring industry but now the curers were finding it very difficult to make a living. Buyers of salted herring would now appear only at the end of the season and they would be looking for bargains. A curer who wanted to stay in the industry had to buy herrings and barrels and hire staff up front, gambling that at the end of the season he would be able to sell all that he had produced.

Bloomfield's premises in Lerwick. On a fine summer day, probably in the 1930s, Bloomfield's most remote yard is a hive of gutting and packing activity with some thirty men and women at work.

The annual trading results during 1920s and 1930s of James More Ltd (part of Mac Fisheries) make grim reading. This had once been a profitable company (Lord Leverhulme did not buy loss-makers) and the fact that they were still in business by 1939 can only be because they were being subsidised by their parent company. What other company could have survived after making a loss in fourteen out of twenty years? The state of the market forced some curers to cut corners in an effort to save on costs but, as we will shortly see, this could backfire on them.

Import duties continued to be a major problem for herring exporters:

Country	Duty (shillings)
Poland	20
Germany	15
Finland	11
USA	10
Lithuania	8
Latvia	7
Estonia	4

Table 13 Import duty on British herring, 1934.

Pressure on Parliament to Help

Throughout the 1920s, the Members for the constituencies in the fishing areas of Scotland waged a campaign of questions and speeches in the House of Commons. They pointed out the hardships of the Scottish fishermen and asked the government to assist the herring industry as, for example, in the aftermath of the Armistice Gale. Because there was no point in catching herrings when you could not sell them, they also called on the government to do more to facilitate trade with Russia.

With the arrival of the 1930s, and with no sign of improvement in the economic situation, they continued to maintain this pressure in Parliament. Their campaign was led now by Robert Boothby, the unconventional Conservative who represented Aberdeenshire and Kincardine Eastern, the constituency that included the major herring ports of Peterhead and Fraserburgh.

HERRING PEOPLE: Robert Boothby

This career politician was a Scotsman representing a Scottish constituency. He frequently got himself into awkward situations in both his Parliamentary career and in his private life (a love affair with Dorothy, wife of Harold MacMillan, for example) but he was one of the most hard-working constituency MPs of his generation. He spent a lot of time with the Aberdeenshire fishermen, gaining a formidable understanding of their industry and of the

Robert Boothby, MP. Robert Boothby was a rather unconventional politician, best remembered nowadays for his sometimes lurid private life. However, he was an untiring supporter of the Scottish herring fishermen and in the 1920s and 1930s tried to sell herrings to Russia.

international herring trade, and in doing so he earned their respect and admiration.

Hansard records that during the early 1930s he spoke as a backbencher 138 times on the subject of the herring industry. When he spoke, he often applied a bit of emotional blackmail by reminding the House of the sacrifices the driftermen had willingly made during the war. His own attempts to boost herring sales included trying to get kippers placed on the approved list of British army rations and making personal visits to Moscow in 1926 and 1934 to try to sell Scottish herrings to the Commissar for Trade (none of these efforts was successful).

The Government acts

Eventually the government decided that it had to be seen to do something. On 20 December 1933, Ramsay MacDonald's National Government appointed the Sea-Fish Commission to investigate the state of the entire British fishing industry, starting with the herring fishery. Ramsay MacDonald had grown up in Lossiemouth (next-door to Hopeman) so he had a good understanding of the problems of the Scottish fishermen. He probably had an even better understanding after visiting Lossiemouth on holiday in 1934 and being 'door-stepped' by a group of Moray fishermen keen to tell him of their plight.

Steam drifters as far as the eye can see! This peaceful view at Yarmouth in November 1933 was taken on a Sunday (the Scots fishermen would not fish on the Sabbath).

The Commission gathered evidence from fishing companies and a wide range of trade organisations, such as the Scottish Herring Producers Association, the Yarmouth and Lowestoft Kipperers Committee, the Rope, Twine and Net Manufacturers' Federation and the Papa Westray Illegal Trawling Committee. Representatives from the Commission were sent around the country to gather information first-hand at fishing ports from Shetland to Cornwall. Bloomfield's Limited was asked to provide the Commission with information about the number of boats they owned, the most efficient type of vessel, the number of staff employed, types of fishing carried out, curing costs, volume of herrings sold abroad, which countries they sold to, the shipping arrangements and the credit arrangements used by other countries. The Commission clearly wanted a full understanding of all aspects of the herring industry and of the problems that each faced.

While this research was going on, a joint approach by the Scottish Herring Producers and the English Herring Catchers Association in 1934 persuaded the government to create a £50,000 fund for the purpose of making loans to fishermen to enable them to buy new drift nets.

The results of the Sea Fish Commission's investigations led to a proposal that the government should take on the role of leading and co-ordinating the herring industry. This would not be nationalisation because the individual companies and employers within the industry retained their independence. Given the fractured nature of the industry, with its numerous trade bodies (boat owners, fishermen, salesmen, curers and so on), this arrangement was seen as the best solution available.

The Herring Industry Board

In introducing The Herring Industry Act of 1935 to the House of Commons, the Minister of Agriculture said: 'The herring industry has been in a state of grave crisis for longer than any of us here care to look back upon.' During the debate it was pointed out that 6 per cent of the herring drifters were 10 years old, 28 per cent between 10 and 20 years old, and 63 per cent between 20 and 30 years old. Some, it was said, were barely seaworthy.

The Act established the Herring Industry Board, which was tasked with 'the re-organisation, development and regulation of the herring industry'. The board itself comprised eight members, three of whom, including the chairman, were to be unconnected with the herring industry. Members from within the industry included Major George F. Spashett (from Small & Co of Lowestoft) and Neil McKay (from Bloomfield's) and representatives of the kippering trade, the curers and the herring exporters. The chairman of the Board would receive an annual salary of £500 (£75,000 at today's earnings) and all other members £250 (£37,500). At the board's first meeting, the deep divisions in the industry were laid bare when the Scots members claimed that the still large Scottish industry was under-represented. It was agreed that more members from Scotland should be appointed to the board. The chairman, James Watt, then resigned on the grounds of ill-health, to be replaced by another Scot, Sir Thomas Barnby Whitson.

Statistics published by the Board showed that in 1913 the average drifter had landed 1,618 crans during the year but in 1934, despite a 33 per cent fall in the number of boats and a 20-year advance in technology, the average catch per drifter was 1,308 crans – 20 per cent lower than 20 years earlier. Such figures did not paint a picture of a modern, well-run industry. There were too many drifters around and too many fishermen struggling to make a living in a much reduced market. By quoting these statistics, the board seems to imply that its main aim was to help the herring fishermen; assisting the entire herring industry was a means to achieving this. Figures for 1935 showed that during the East Anglian season the average total landings by English boats was 1,100 crans and by Scots boats 500 crans, proof of the poor state of the Scots fleet.

The Hopeman Herring Fleet, 1936. This image of Hopeman harbour shows that, despite some eighteen years of recession in the herring industry, there were still at least seventeen steam drifters owned by local families.

What were the Responsibilities of the Board?

The board's remit included:

Promoting herring sales, market development and research
Making loans for the construction, re-conditioning and equipping of boats
Making loans to assist in the export of herrings
The purchase and disposal of redundant boats
Limiting the numbers of boats, curers, salesmen, kipperers, processors and exporters by means of a licensing system
Regulating the industry by means of prohibitions and restrictions with a view to meeting temporary and seasonal conditions detrimental to the industry
Acting as an agent in the purchase and sale of herrings for export, including the power of compulsory purchase
Disposing of surplus herrings by conversion into oil and other products
Obliging all in the industry to keep accurate records, accounts and to provide the board with information

This comprehensive list covered most aspects of the industry. The sting in the tail was that the board's services were not to be provided at the tax-payer's expense. They were to be funded mainly by a series of levies and licences which all participants in the herring industry would have to pay. In its first year the board raised almost £2,000 by means of these license fees.

In addition to the licenses there was a levy on all herring landed in the United Kingdom of one (old) penny for each pound worth sold – in 1936 this brought in £11,500. Anyone within the industry who failed to fully comply with the board's regulations could be fined £5 for a first misdemeanour and up to £20 on any subsequent occasion.

Type of License	Price			Basis of charge	Period of License
	£	s	d		
Drifter	-	2	6	Vessel, over 20 ft length	4 months
	-	10	-	Vessel, over 60 ft length	4 months
Herring Salesmen	1	-	-	For each port worked in	One year
Herring Curer		5	-	For each port worked in	One year
Kipperer	1	-	-	For each port worked in	One year
Herring Exporter	2	-	-	First 5000 barrels exported	One year
	-	2	-	Each additional 1000 barrels	One year
	1	-	-	Additional license for Klondyke trade	One year

Table 14 Herring Board license fees.

One notable omission from the board's role was any mention of conservation or management of fish stocks. Current scientific research into the herring was still directed towards increasing catches.

Progress by the Board

In 1936 the board spent £25,000 on national advertisements within the United Kingdom to promote sales of fresh herrings. A series of lectures and herring cookery demonstrations were provided to organisations such as Women's Institutes and at events like the Ideal Home Exhibition. They also produced *The Herring Book*, which contained articles on the health benefits of eating herring and featuring herring recipes. This was for distribution free from fishmongers.

The board were confident that this promotional activity had increased the home demand for herring but unfortunately the 1936 herring catch was a poor one and prices rose, leading some of the public to believe that the price had been deliberately raised as part of the sales push! Undaunted, in 1937 the board produced *The New Herring Book*, which was packed with more recipes – who could possibly resist 'Kipper Scramble' or 'Herrings in Cream Sauce'?

With regard to foreign markets, the board's promotional efforts created interest from Australia, South Africa and Palestine. Sales efforts in the Middle East, however, were disrupted by 'disturbances between Jews and Arabs'. More significantly, the board sent a delegate to Russia to try to revive the Russian market.

As for the fishing fleet, in 1936 the board advanced over £32,000 to fishermen in loans for the purchase of new drift nets. They also made progress on reducing the number of drifters by buying 113 vessels at an average of £115 each. They then disposed of these to breakers' yards for £50 each. This may sound expensive but if it encouraged more fishermen to leave the industry, it would become more efficient – by 1938 the Scottish herring fleet was half the number it had been in 1920 (Table 15).

The Herring Industry Board produced this new edition of *The Herring Book* in 1937, with more recipes and a celebrity endorsement from film star Merle Oberon.

Importantly, the Board tried to prevent the herring gluts which had previously ruined market prices and had led to perfectly edible herrings being used as manure. They appointed an official to each herring port to monitor landings and sales. If there were indications that the curers and other buyers were struggling to cope with the volume of herrings, he would intervene by preventing the boats from returning to sea until the markets had cleared.

Firm start and finish dates were set for herring seasons; in 1935, for example, in Orkney and Shetland no more curing was allowed after 2 September and at Yarmouth curing was not allowed to start until 8 October. The Board, recognising that the basic problem was over-supply of salted herring, was attempting to both limit the amount on the market and to ensure that what was available was all good quality, produced from prime fish.

A Difficult Balancing Act

While the Board was there to lead the industry, the industry itself was still composed of independent companies and individuals, grouped together in their trade associations: the Herring Salesmen's Association; the British Herring Trade Association; the Great Yarmouth Mediterranean Exporters Association; and so on. Their individualism sometimes caused the Board a degree of frustration. In 1936, while the Board was on a sales drive in Russia it found that the Russian negotiators absolutely refused to pay any more than 27 shillings and 6 pence per barrel. The curers protested that this price did not even cover their production costs and so the Board abandoned the negotiations, only to hear two months later that one of the voices of protest had just sold 30,000 barrels to the Russians at that price.

In 1937, having gained a contract from Russia for 70,000 barrels, the Board negotiated a loan from the National Bank of Scotland to the curers to help them finance the completion of this order. The bank wanted to make the money available as a single loan but the British Herring Trade Association stalled negotiations by insisting on individual loans to all 130 of its members. It seemed that for all its apparent authority, the Board lacked the power to knock heads together.

In 1938 the Board approached the Nederlandsche Visschericentrale with a proposal that the British and Dutch herring trades should combine in a joint marketing venture directed particularly towards Russia. The Dutch response was that this was 'not regarded as desirable by the Dutch industry' – another setback for the Board.

A further hurdle for the Board was that, although it had the role of managing the herring industry for the whole of the United Kingdom, it had to deal with two separate government departments – in Scotland the Scottish Office (Edinburgh) and in England and Wales the Ministry of Agriculture and Fisheries.

Contraction of the Herring Industry in Scotland

Table 15 confirms that as the 1930s progressed, fewer Scots were relying on the old technology of steam power.

Year	Number	Yearly Reduction
1930	802	-
1931	799	3
1932	770	29
1933	733	37
1934	686	47
1935	644	42
1936	536	108
1937	459	77
1938	402	57

Table 15 Scottish herring drifter numbers, 1930s.

Further evidence that the Scottish herring industry was contracting can be seen in the numbers of curers and herring salesmen working the summer fishing at Wick (Table 16). Andrew Bremner would have been turning in his grave!

Year	Number
1930	60
1931	58
1932	54
1933	54
1934	46
1935	40
1936	36
1937	32
1938	30

Table 16 Herring curers at Wick, 1930s.

Despite these falling numbers, the herring industry was still integral to many Scottish communities, to the extent that in the 1930s both Wick and Eyemouth introduced into their summer gala celebrations a pageant in which each year a local young woman was elected 'Herring Queen'.

Even though the numbers of boats and curers in Scotland was falling, the fishing incomes were still very low, leading to fishermen at Fraserburgh and Peterhead going on strike at the start of the 1936 Scottish herring season. They wanted a guaranteed income of 30 shillings (£1.50) per week. They got what they wanted by all branches of the local herring trade (boats, merchants, curers, exporters, etc.) agreeing to pay a levy (2 pence in the pound) on all their transactions during the season, the money collected to be placed into a reserve ready to pay the fishermen should it be necessary.

In the 1930s, one of the concerns regarding the herring industry was the low volume of herrings being bought by the British public. It was thought that the main reasons for this were, firstly, snobbery (they were viewed as a cheap fish) and, secondly, the bones – if you bought filleted whitefish you did not have to cope with fish bones. In 1930 it was estimated that only 10 per cent of the British herring catch went to the home market. In 1931, Miss Thursby Pelham wrote to the *Lowestoft Journal* with a suggestion for boosting the herring trade. She said: 'Why not eat salt herrings in this country? We have all got too used to tinned salmon from America!'

Herring People: Dorothy Thursby-Pelham

Miss Thursby-Pelham was a scientist who, between the wars, worked at the Lowestoft fishery research laboratory. She had the distinction of being the first female scientist to regularly go to sea – for a week at a time – on a fisheries research vessel. In 1937 she delivered a talk on BBC Radio on her experiences of working at sea.

HERRING PEOPLE: Simone Barnagaud-Prunier

Simone Prunier was born in 1904 into a dynasty which owned a famous Paris restaurant and fish supply firm. In 1920 she joined the family firm and in 1922 married Jean Barnagaud. Together they ran the family business after the death of her father. In 1936 it was decided to open a branch of Maison Prunier in St James, at the heart of London's West End. Simone was despatched to set up the new business and ended up staying for many years.

Not long after opening the restaurant, she got into a discussion about ways to promote British herring. Her idea was to offer an annual trophy to the boat that made the largest overnight catch during the course of a season. The (not

Above left: Madame Prunier. Simone Prunier was born into a restaurant dynasty in Paris. In 1934 she left Paris to open a London branch of the family restaurant and in 1936 became involved in promoting British herrings.

Above right: The Prunier Trophy. The winning skipper was allowed to keep the Prunier Trophy for a year. It was carved out of Purbeck marble by sculptor Charles Sykes.

entirely logical) thinking behind this was that the media publicity generated by the competition would encourage the British public to eat more herrings. Her suggestion gained the approval of both the Ministry of Agriculture and Fisheries and the newly-constituted Herring Industry Board. A committee of advisers was appointed (which included Robert Boothby) and the first competition was held in 1936. It became an annual event and it ensured that the name Prunier became permanently associated with the British herring industry.

Trophies and Prizes

The competition for the Prunier Trophy was held annually during the East Anglian season and it ran from 5 October to 28 November. Because the majority of boats were Scottish, if the first prize (a sculpted trophy and £25) was won by an English vessel, then the second prize (also £25) would go a Scots vessel and vice-versa. The rules stipulated that the winning catch should be from just one night's fishing.

Some astounding overnight catches were recorded during the years of the Trophy – we will see the story one of them in Chapter 9.

Competitions were not new; some English fishing companies were already running their own in-house herring catching competitions. Jack Breacher Ltd of Lowestoft, for example, awarded a first prize of £40 to whichever of their boats made the best annual earnings and also gave £10 to the best maintained boat (at a time when their bank overdraft was nudging £50,000). Bloomfield's, too, ran an annual competition, the D'Arcy Cooper Challenge Cup, which was named after Lord Leverhulme's successor as chairman of Lever Bros. This competition offered prizes to the highest earning Bloomfield vessels on the Home Voyage; £100 went to the top steel-built boat (won in 1933 by the drifter/trawler *D'Arcy Cooper*!) and £100 to the top wood-built vessel (plus a second prize of £50). These prizes were split among the crew according to the crews' shares, meaning that the winning skipper received about £17 and the cook just over £7. In 1933, the drifter *D'Arcy Cooper* also received a £24 'Skipper's Bonus' from Bloomfield's for achieving over £4,500 in earnings for the year. So, cash prizes were not by any means new to the industry but the Prunier Trophy differed in that it was open to all vessels, regardless of who owned them.

The Prunier Trophy ran till 1966. It was successful in attracting public and media interest and it generated keen competition among fishermen. However, there is no evidence that it led to any increase in home sales or that it was as effective as the advertising campaign run by the Herring Industry Board. If the market is already struggling to sell herrings, will encouraging more to be caught lead to an increase in sales? What did it do to improve the quality of the product and did it address the questions of consumer resistance – snobbery and coping with the bones?

Technology and Research

In the 1930s, early experiments with sonar were refined and developed into equipment that could help fishermen detect shoals of fish. In 1933 echo sounders were fitted to a number of trawlers and proved to be successful. This new technology was adopted by Bloomfield's in 1934 when they signed a contract with Marconi Sounding Device Co. Ltd to fit their drifter/trawlers with the 'North Sea' type Echometer Sounding Device. The cost for this was £57 10s per vessel. At that price echo sounders were not affordable to most independent boatowners – for some of them, the equipment would have cost more than their boat was worth. By 1935 some trawlers and a few drifter/trawlers were being fitted with radio to assist ship to shore communication.

The fishery research laboratories continued to investigate the herring stocks and their migrations. For several years, researchers met each boat landing at Lowestoft and Yarmouth to take details of the catches and to ask the skippers where they had made their catch. These details were matched against weather records to see if the speed or direction of the wind had any effect on the migratory route of the herring.

Research was carried out during the 1920s and 1930s to see if there was any truth in the old fishermen's belief that the best catches were made at full moon. It turned out to be true, but only when the full moon fell during the second week of October.

The best way to trace the herrings' migratory routes was to fix a tag to newly-caught fish and then return them to the sea alive in the hope that the next time they were caught, someone would report back the details of where and when. The information so gathered provided clues as to what distances herrings swam, their routes and their speed. It proved, for example, that the herrings caught in summer off the Yorkshire coast spent some of the year off the coast of Denmark.

In the 1930s the industry discovered the convenience of bulk freezing (something that Lord Leverhulme had proposed way back in 1919). At Yarmouth in 1935 and again in 1936, a refrigerated cargo ship, the *Thorland*, was used to freeze some 5,000 tons of herring. These were taken to cold storage in Hull, to be gradually released to smokehouses in the spring when there were few fresh herring around.

In 1938 the national press reported on the experimental use of an aeroplane in the Shetlands to search for shoals of herring (something else that Lord Leverhulme had proposed!). Drifters wasted a lot of time and fuel in searching for herring shoals – this unproductive cost had been calculated at £200 per boat per season. It was found that spotting shoals from the air saved time and could cover a much larger area of sea. Because many drifters did not carry two-way radios, once the pilot spotted a shoal he would have to drop a message to the fishing boats. The drifters that followed his directions consistently made large catches of herring. The Herring Industry Board would have continued the experiment the following year had not war intervened.

Selling Herrings Overseas in the 1930s

From the early 1930s the German government imposed strict *devisen* or currency restrictions that made it harder to export goods to Germany. When the Nazi party came to power in 1934, they re-introduced these restrictions, causing more difficulties for the British herring curers and exporters.

In 1934, Britain and Latvia signed a trade agreement under which Latvia would buy 10,000 tons of British cured herring and impose only a low rate of import duty on them. In the few weeks that it took to deliver the herrings, a new Latvian government came to power and they immediately raised the import duty. Bloomfield's increasingly frustrated manager in Latvia asked if the Yarmouth Head Office could enlist the help of their parent company, Unilever Ltd. Under the 1934 trade agreement, Unilever were importing large quantities of butter, eggs and bacon from Latvia. Could they not, he asked, use their influence to make the Latvian government buy more herring from Bloomfield's? Sadly, they could not.

In the Baltic countries, the British herring sellers still faced strong competition from the Dutch and the Norwegians. Strict attention had to be paid to the quality of the product and to the after-sales service. In 1936 and in again in 1938 Bloomfield's had to send representatives out to Klaipeda (Lithuania) and to Gdynia (Poland) to deal with complaints about the quality of herrings that they had sold. There were particular problems with the 1938 Stronsay herrings, produced by their subsidiary company, Joseph Slater Ltd. Problems included thin and torn herrings, poor gutting (40 per cent of the herrings, it appeared, had not been gutted properly) and weakness of the pickle (the brine solution). After investigation, some of the problems were attributed to the curer's practice of using the more expensive Torrevieja salt at the top and bottom of the barrels but using cheaper Cadiz salt in the middle. The curer had clearly been cutting corners in order to keep costs low.

In 1938, Bloomfield's experienced further problems in Klaipeda (also known as Memel) when the Lithuanian government allowed a total import quota of just 60,000 barrels, of which Bloomfield's were only allocated 5 per cent. Their previous market share indicated that they should have been allowed at least 20 per cent of the quota. Things took a more sinister turn that year when their branch manager in Klaipeda, Herr Elsner, a German from Danzig, was (falsely) accused of being a Nazi by the (mainly Jewish) local herring importers, who then refused to handle any Bloomfield's herrings.

A condition of the new trade agreements with Russia was that the United Kingdom should import salt from Russia to use on the salted herrings sold to Russia. When Bloomfield's tested the Russian salt, they found it left a dirty chalky grit at the bottom of the barrels of herring and instead of being a rich brown colour, the pickle was a milky-white. If they had used this salt they would have received more complaints regarding quality.

HERRING PEOPLE: Tom Bruce

In 1932 Tom Bruce and his partners sold the *Admiration* to the Main family of Hopeman and left herring fishing. Tom and his friend James Sutherland jointly invested in a motor fishing vessel, the *Prevail* (INS 308), and like other former herring fishermen he took up seine netting for haddock off the north and west of Scotland. As soon as his sons left school, they joined him on the *Prevail*.

The arrival of the *Prevail* brought about a change in the fortunes of the Bruce family. Tom was still in debt but was now, in financial terms, beginning to get his head above water. It was not long before the family moved into a house of their own where the boys could at last sleep under the same roof as the rest of the family.

HERRING PEOPLE: Wee Green

On 10 November 1931, one of Wee's drifters, the *Ocean Swell*, was fishing off Smiths Knoll when the crew spotted a disabled cargo vessel. They went to

BLOOMFIELD'S OVERSEAS, LTD., MARSTALL STRASSE, 19 RIGA. TELEGRAMS: HERINSELER.	TELEGRAPHIC ADDRESS: HERINSELER, GT. YARMOUTH. TELEPHONE: 691 (5 LINES).	BOLMA HERINGSHANDELS G.M.B.H. HOLZSTR, 1, STETTIN. TELEGRAMS: HERINSELER.

BLOOMFIELD'S OVERSEAS, LTD.

HERRING EXPORTERS & SHIPBROKERS.

REGISTERED OFFICE:

GREAT YARMOUTH.

AND AT

GLASGOW - STORNOWAY - LERWICK - STRONSAY

WICK - FRASERBURGH - PETERHEAD - ABERDEEN

LOWESTOFT.

OCEAN HOUSE,

GREAT YARMOUTH.

BLOOMFIELD'S OVERSEAS, LTD., SAEGERPLATZ, 16, LIBAU. TELEGRAMS: HERINSELER.

BLOOMFIELD'S OVERSEAS G.M.B.H MUNCHENGASSE, 4/6 PT, DANZIG. TELEGRAMS: HERINSELER.

BLOOMFIELD'S SP. AKC. GDYNIA. TELEGRAMS: HERINSELER.

BOLMA HERINGSHANDELS G.M.B.H. DIRCKSENSTRASSE, 47, BERLIN, C.25. TELEGRAMS: HERINSELER.

F. BOEHM & CO., G.M.B.H. DOVENHOF, HAMBURG, 8. TELEGRAMS: HERINGBOEHM.

BLOOMFIELD'S OVERSEAS (ALTONA) G.M.B.H. PALMAILLE 29, ALTONA, ELBE. TELEGRAMS: HERINSELER.

Lowestoft, 11th. December 1937.

Bloomfield's Overseas Ltd was the export arm of Bloomfield's Limited. This headed paper illustrates both the national and the international nature of the herring business.

EL   00647

HERRING INDUSTRY SCHEME, 1935.

EXPORTER'S LICENCE

MESSRS. BLOOMFIELDS OVERSEAS, LTD. .... of OCEAN HOUSE, GREAT YARMOUTH

is hereby authorised to act as an exporter of (1) fresh herring (2) and herring other than fresh herring during the year 193**8**, subject to the following conditions:—

1. The Licensee shall act in conformity with all relevant provisions of the Herring Industry Scheme, 1935, and with all relevant rules, requirements and directions, from time to time made or given thereunder by the Herring Industry Board or with their authority.

2. The Licensee shall, if required by the Industry Board, furnish to the Board or such person as they may appoint, a statement with regard to each shipment made by him, showing the quantity and selections exported and the ports from and to which they were sent.

3. This licence shall be produced on demand to any member or officer of the Herring Industry Board or to any person duly authorised by them and if so required by them shall be surrendered to them for suspension or cancellation.

4. If the holder of this licence satisfies the Herring Industry Board that it has been lost or destroyed a new licence will be issued on payment of a fee of 2s. 6d.

By order of the Herring Industry Board,

Dated *12 . 5 . 38*.

FEE OF RECEIVED.

Secretary.

Herring Export License. This license issued by the Herring Industry Board permitted Bloomfield's to export herrings.

investigate and found that the Dutch coaster *Berent* had lost all engine power. The *Ocean Swell* towed her to Yarmouth harbour entrance, where she was handed over to a tug which took her into the harbour for repairs. Wee registered a successful salvage claim for £500 (£75,000 at today's values).

At the outbreak of the Second World War Wee's son, Vernon, recalls seeing his father being chauffeured away in an Admiralty car for further talks on anti submarine nets (which Wee had been instrumental in developing during the First World War).

Timelines Chapter 6

1930 The steam drifter/trawler *Lydia Eva* is built at Kings Lynn

1932 *Wilson Line* is the last steam drifter to be built in the UK (by Hall & Co., Aberdeen)

1933 Trawlers start to be fitted with echo sounders to help find fish shoals

1933 The government appoints the Sea Fish Commission

1934 Robert Boothby MP visits Moscow to personally promote British herrings

1934 The Sea Fish Commission report on the herring industry is published

1934 Bloomfield's fits echo sounders to its drifters

1934 New herring trade agreement signed between UK and Latvia

1935 The Herring Industry Board is established

1935 Early efforts to bulk freeze herring using SS *Thorland*

1936 The Herring Industry Board prints *The Herring Book*

1936 The Prunier Trophy is launched

1936 113 steam drifters are scrapped by Herring Industry Board

1937 *The New Herring Book* is published by the Board

1938 Experimental use of aeroplanes in Shetland to detect herring shoals from the air

1939 The Danes begin industrial fishing of herring and other fish

CHAPTER 7

The Herring Girls

Beginnings

In coastal communities, women had always assisted the men by mending nets, gutting fish and sometimes even carrying the fishermen on their shoulders out through the surf to the boats. The phenomenon of a large female workforce travelling far from home was very different and it had its beginnings in nineteenth-century Scotland, at Wick, when young women from the Hebrides walked across Scotland to work at the summer herring season. In Wick they lived in unhealthy conditions in an assortment of sheds, bothies and huts, surviving several outbreaks of cholera that occurred there during the Victorian era. As the industry grew, the Hebridean women were joined by girls from herring ports in Moray, Banff, Aberdeenshire and Fife. At the end of the summer season the women were paid off and they made their way back home.

When, in the 1870s, the Scottish curers spotted the potential of working on the east coast of England, they wanted to have an experienced workforce with them. As the English were unfamiliar with the Scotch Cure and Crown Brand requirements, it seemed sensible for the curers to bring workers from Scotland. By the 1880s Scottish herring girls, many of whom spoke only Gaelic, were a familiar sight to the residents of North Shields, Scarborough, Yarmouth and Lowestoft. Whatever their age, these women were known in England as 'herring girls' or 'fisher girls'.

It soon became an annual routine for Scots girls to travel north in the summer and south in the autumn, following the various herring seasons. In summer they worked at Wick or at herring stations in Stornoway, Orkney and Shetland. Some would then work at the Fraserburgh and Peterhead seasons. In late summer they would go south to Northumberland and Yorkshire and, at the end of September, to East Anglia. Wherever the Scots curers set up in business, including the Isle of Man and Northern Ireland, they took Scots girls with them.

Hopeman girls at Stromness in 1906. These Hopeman herring girls had put on their best clothes to have a souvenir picture taken in Stromness, Orkney. The girl on the far left of the middle row is Jessie McPherson, who in 1913 married Tom Bruce.

Fishworker's train ticket. At the height of the herring industry the railway companies pre-printed stocks of tickets specifically for use by the herring girls as they travelled between Scotland and East Anglia.

Recruitment

Young women were mostly recruited from the coastal villages around northern Scotland and the Hebrides but there were also a few from Ireland. At the turn of the century, the career choices for young women from these areas were severely limited – they either joined the herring industry or went into 'service' as a housemaid, probably in one of the big cities. Gutting herrings was not pleasant work but the pay was better than that of a housemaid and it meant that one saw more of one's home and family.

In the early 1900s the girls were mostly young and single, aged from fourteen upwards. In the 1920s the average age of the workforce began to creep higher as married women returned to the job having brought up their families, or having been widowed during the Great War. By the late 1930s the herring industry was contracting so fewer young girls followed their mothers and aunts into the herring gutting, which again caused the average age of the 'girls' to rise.

Before each season, the curing companies sent agents round the fishing villages to engage crews (two gutters and one packer formed a 'crew') and to agree pay rates. On signing up for the curer, the girls would receive an initial engagement fee called 'arles' (from the Gaelic *a earlais*). This fee was generally 10 shillings (50p) and in accepting it the crew committed themselves to stay with that curer for the season. The girls each packed a kist (a wooden chest or trunk) with work wear, best clothing, foodstuffs and other items (see Appendix 5) that they would need while away from home for eight to ten weeks.

A herring girl's 'kist'. A 'kist' was a trunk in which each girl would pack all she would need for a two-month absence from home. They were often made by a local carpenter.

Travel and Accommodation

The curers arranged transport for the girls, getting them tickets on scheduled steamers and ferries to the northern isles for the summer season. For the East Anglian season, they chartered special trains departing from Fraserburgh, Oban, Mallaig, Kyle of Lochalsh and Aberdeen. The curers' male staff – coopers, apprentices and labourers – loaded the heavy kists into the luggage vans and the journey south began. Typically, these trains ran overnight (it took some fifteen hours to reach Yarmouth from Aberdeen), having taken a circuitous route across the Fens and around rural North Norfolk via the Midland and Great Northern Railway. At the beginning of each herring season, Yarmouth's Beach Station would receive as many as twenty of these trains in a week (sometimes six in a single morning), each one carrying 200–300 workers and their belongings. Once the girls for Yarmouth had alighted, the trains travelled forward to Gorleston and Lowestoft, taking the girls who were signed up to work at curing yards in those towns. On arrival at their station, the girls and their kists would be transferred by curer's lorry to the digs that the curers had arranged for them.

The girls would have been grateful for the efforts made on their behalf in the early 1900s by Alfred Bloxham, the station master at Yarmouth Beach Station. He had not been very impressed with the outdated and very basic carriages that were being used for the herring worker specials and in early 1905 he travelled north to enlist the co-operation of the Scottish railway companies. From that year onwards, thanks to his efforts, the girls made the long journey south in more modern corridor coaches which now included toilet compartments.

In England the girls generally lodged with a family. By our standards, the accommodation was basic and cramped – up to three girls shared a bed in a room

Accommodation in Yarmouth. The 'digs' provided for the Scottish herring girls in England were quite basic. Three girls were accommodated in the front bedroom of this little terraced house in Yarmouth each herring season up until the late 1950s.

that had been stripped of all decent furniture and floor coverings (landladies were all too aware of how long the smell of herrings could linger in soft furnishings). It was warmer and more secure than the wooden huts that were provided for them in the more distant corners of Orkney and Shetland. In Yarmouth and Lowestoft some landladies would cook meals for them (the girls bought the ingredients) but in Orkney and Shetland they had to cater for themselves.

The Working Day

At quarter to seven in the morning, the girls put on their rubber boots and bibbed oilskin aprons (which their landladies insisted were kept outside in the back yard) and the curer's lorry ferried them to the curing yard. To protect their fingers from the sharp knives, they wrapped them in 'cloots', finger bandages made from strips of clean cotton or linen (old flour bags were a favoured source of material) and tied on each finger with string.

Soon the first cartload of herring arrived from the fish quay and was tipped into the farlans. The first stage was for the coopers to 'rouse' the herring (shower them with coarse salt and then mix it in). The salt helped prevent any decay of the fish and it also made them less slippery to pick up by hand. The girls stood in the open air at the farlans and started gutting or 'gypping' the herring with a small sharp knife; one small incision, a quick twist and the gut and gills were out and dropped into a small basket and the gutted fish was thrown into one of several tubs behind the girls (one tub for each Crown Brand size). The herring guts would usually be collected by local farmers for use as manure. Experienced girls worked at great speed and could consistently gut between thirty and sixty herrings per minute.

When the tubs were full, the two gutters carried them to the back of the plot, where their packer transferred the fish into barrels, lining them neatly in layers with a layer of salt in between. The tallest girl in the crew was normally the packer because a shorter girl would have difficulty reaching to the bottom of a barrel to arrange the first layers of herrings. Sometimes the packer would be the left-hander of the trio because with the sharp blades flashing around at the farlans, it could be quite hazardous having a left-hander in among all the right-handers. Each barrel held roughly a cran (over three hundredweight) of herring together with a hundredweight of salt. A cran, depending on the size of the fish, comprised 800 to 1,200 fish. Sometimes the curer produced smaller half-barrels but most customers wanted full-sized barrels. In 1924, one crew managed to fill thirty-seven barrels in a day, meaning that the two gutters in the crew must have each gutted over 18,500 fish. As each barrel was filled, the coopers placed a weighted lid on top, chalked the date on the side and left it to allow the contents to settle for a few days.

It was outdoor work and the girls stood all day in a mire of mud, sand and herring offal. The work continued all day, with a couple of meal breaks, until all the herrings that the curer had purchased were barrelled. When herring were in good supply, the

Herring girls, 1900. This image from Yarmouth shows how low the farlans originally were. In later years they were raised on wooden blocks in order to put less strain on the backs of the herring girls.

girls routinely worked until six one night and nine the next. If there was a glut of herrings, they worked till very late. When darkness fell the coopers rigged up lines of oil lamps or naptha flares so the work could continue. In Yarmouth the fish auctioneers ran four sales sessions each day, the first starting at 7.00 a.m. and the last finishing at 9.00 p.m. This meant that at any stage between those times, curers could move into the market to buy herring and would need the girls to process them immediately. If it started to rain or snow (the East Anglian season lasted till early December), the coopers hoisted tarpaulins which offered some protection from the worst of the weather but not from the cold. Facilities at the pickling plots were very basic. At Yarmouth in 1903 the Factories Inspector had to badger the Council into installing flush toilets on the pickling plots; sixteen were built to serve some 2,000 women. Before 1884 there had been no toilets all for the women until a local herring smoker, Mr C. Stacy-Watson, prevailed upon the Council to erect some 'in the interests of decency and morality'.

The Factory and Workshop Act (1871) and HM Inspector of Factories Act (1898) set maximum working hours for women which applied to most industries. Herring curers, however, were exempted from both these pieces of legislation so the girls sometimes had to work into the early hours of the morning.

Topping Up

On Mondays, only a few boats landed herring, so that day was known in the curing yards as 'topping up day'. After about ten days, the herrings in the barrels had shrunk as the salt drew the blood and oil from them, forming a dark brine. Because of the shrinkage, the barrels now needed to be topped up before they could be finally closed up and made airtight.

First, the brine or pickling liquor was drained off into a bucket. It was important not to lose any of it because it formed part of the gastronomic experience for the European consumer. To compensate for the shrinkage, a further two layers of herrings were placed in the barrel – it was important that these herrings came from the same date-batch as those they were joining. The barrel lid was then sealed and the original liquor was poured back with a funnel through the bung-hole, which was then plugged. The barrel was now ready for a final check by the coopers, and to be 'branded' with the date and the curer's name. The brand was applied by means of brushing tar across a metal stencil.

Pay

The girls received a weekly housekeeping allowance which covered their food and accommodation and at the end of the season they were paid for the number of barrels that they had filled. By the 1920s and 1930s, the average rates of pay at the East Anglian season were:

Arles	10s (50p) each
Weekly living allowance	15s (75p) per crew
Topping up	6d (2½p) per hour each
Barrel money ('piece rate')	1s (5p) per barrel (to the crew)

Table 17 Herring Girls' pay rates in East Anglia, 1920–38.

If landings were poor, there was no work for the girls and, therefore, no pay. At these times they knitted, even as they strolled around town. If things were really quiet in the curing yards, some girls would look for a day's work splitting and cleaning herring in one of the smoke houses.

Their total pay for a season varied considerably, depending on how quickly they worked and how good the landings had been, and it is not easy to find accurate details of the pay that individuals received. In 1911, it was reported that the girls

Topping up
These two girls are not handling salt nor are they wearing 'cloots', so we can deduce that
they are engaged in 'topping up' barrels after the initial shrinkage of the herrings.

were each taking home an average of £16 at the end of the Yarmouth season. At
the end of the 1924 season, the chairman of Mac Fisheries, Sir Herbert Morgan,
told *The Times* that his gutting crews had earned £90 for the ten-week season – he
may not have realised that that money was not per person but was shared between
three girls and that very few crews earned that much.

The Number of Girls

How many herring girls were there? Employment was seasonal and short-term so
it is difficult to find accurate figures. There were numerous estimates and among
the most reliable are the following:

In 1883 it was estimated that 300 Scots girls were working in the curing yards
during the Yarmouth season
In 1912, the Scottish Office reported that 10,800 women in Scotland relied on the
herring industry for their living

In 1913, 2,400 girls from the Isle of Lewis worked the Shetland season and 1,600
  of them later went to Yarmouth and Lowestoft

In autumn 1931, the railways carried 5,133 girls from Scotland to Yarmouth and
  Lowestoft

In 1939, over 6,000 women worked as gutters/packers at the Scottish summer
  herring fishing

Leisure Time

Any free time that the girls had might be spent washing and mending their clothes
or socialising, perhaps meeting up with friends from their home villages. In Great
Yarmouth and Lowestoft, the girls were drawn to the town centres – the variety
of shops and the choice of consumer goods were far better than that in their home
areas. They spent a lot of time window-shopping and noting what they would
buy when they got their barrel money. Shopkeepers were generally quite good at
putting items aside for the girls until they got their barrel money at the end of the
season.

On Sundays most girls attended church. Afterwards, there was a chance to meet
family and friends and maybe to visit the cinema. Being away from home gave the
girls a chance to socialise with young men that they knew from home without being
watched by anxious parents. The girls would sometimes meet up with male family
members (fathers, brothers and cousins) when the Scots boats were in harbour.
Some girls met local men and romances occasionally blossomed, leading to 'mixed
marriages' between the Scots women and English men.

Welfare

From the very early days of herring girls travelling away from home there had
been concern for the girls' physical health and for their moral well-being. When
the curers first took herring girls to England, the Scottish Episcopalian Church sent
a minister and a trained nurse to Yarmouth and to Lowestoft to look after them
during the season. Other Scottish churches and denominations soon followed suit.
Many of the girls had never seen a large town before, especially one with all the
attractions of the English seaside.

In time, various other organisations provided shelter and medical treatment for
the girls. Each season from 1899 until the First World War, a Miss Davidson ran a
rest room on Admiralty Road in Yarmouth where the girls could receive medical
treatment from two trained nurses or just enjoy a cup of tea and some sympathy.
The funding for this establishment came partly from Scottish churches and partly
from philanthropic donations. After the 1911 National Insurance Act was passed,
the women became entitled to free medical care from doctors and hospitals while

in Yarmouth or Lowestoft, but for many of them a rest room remained their first point of medical treatment.

Herring People: Margaret Harker

Margaret Harker of Blofield Hall in Norfolk lived less than half an hour from Yarmouth. During the First World War she had run a fifty-bed Red Cross hospital, and by 1918 she had been promoted to Assistant County Director of the Norfolk Red Cross. In the 1920s she became concerned about the health and welfare of the Scots herring girls. Margaret was a fellow Scot and doubtless felt an affinity with the girls.

At the beginning of the 1929 herring season, she opened a Dressing Station and Rest Room in Yarmouth. Medical care was provided by Red Cross VADs and in the same building the Scottish churches provided social care for the girls. The Dressing Station was very successful and much appreciated by the visiting Scots workforce and it operated each season throughout the 1930s. Margaret and her staff were held in high regard by the herring girls and they received many letters of thanks from Scotland.

Margaret died in early 1935 and the following year her life was commemorated in a series of stained glass windows in Blofield church. This memorial includes two pictures of the herring girls. While there are hundreds of black and white photographs available of the 1930s herring girls, these windows provide rare colour images of the girls from that decade.

Common Injuries

The most common injuries treated by the Red Cross were cuts and salt sores. Despite their protective 'cloots', the girls often sustained small cuts, which as the day progressed were increasingly exposed to fish guts and brine. When this unsavoury mixture penetrated the skin, it could be painful and could cause infections and even blood poisoning. Salt crystals could cut into the softer areas of skin, especially the web between the fingers, and cause painful sores. While they were working, the girls' hands were often so cold that they were not immediately aware of any cuts or abrasions.

Red Cross statistics record that during the course of a season some 20 per cent (over 1,000) of the herring girls in Yarmouth turned up with injuries at the Dressing Station and that many of them had to return daily for up to 10 days to have a fresh dressing applied. A girl would rarely take any time off for illness or injury because to do so reduced the earnings of the other two girls in her crew.

A common mishap in the curing yards was that a herring scale (sometimes the air seemed full of them) could get lodged in your eye. A herring scale defies all

Above left: In 1929 Margaret Harker, head of the Red Cross in Norfolk, opened a dressing station for the herring workers in Yarmouth.

Above right: Window at Blofield Church, Norfolk. This window forms part of the memorial to Margaret Harker.

normal methods of removing a foreign body from the eye but in each yard there would be one woman who knew how to deal with them. She sat the victim down, held her eyelid open and, using only her tongue, located the scale and flicked it clear of the eyeball.

Industrial Disputes

From old photographs and postcards we might think that the girls were always happy and smiling. This may have been the case when someone pointed a camera at them but, as we have seen, they worked very hard in conditions that would not be tolerated nowadays. The smiling exterior masked the inner strength which was needed to do that work. The first recorded instance of their resorting to industrial action was in the Shetlands. At Lerwick, the girls were normally accommodated six to a hut but in 1930 they refused to work unless this number was reduced to a more comfortable three. Their demands were met.

As the Depression of the 1930s added to the financial pressure on the herring trade, one of the curers' cost-cutting measures was to reduce the herring girls' pay rates. In 1935 they cut the barrel money from 1 shilling (5p) to 10 pence (4p) a barrel, a pay cut of 17 per cent. In 1936, before they travelled to Yarmouth, a handful of women decided that they would strike for an extra 2 pence a barrel to bring the rate back to one shilling. On arrival in Yarmouth, they spread the word among like-minded crews to be ready to strike when the time was right. The

In 1929, the newly-opened Red Cross Dressing Station in Yarmouth attracts a lot of attention. Note the 1920s fashions and the very basic staff transport provided for the girls.

The interior of the Red Cross Dressing Station in a calm moment before the doors were opened to patients.

herring season began quietly with landings low and so they bided their time. In the fourth week, when the herring began to appear in great numbers and just when their labour was most needed, they began their strike. The ringleaders quickly moved from yard to yard, persuading or threatening their fellow workers to join them.

They were not members of a trade union and so lacked the organisational backing and skills that a union had at its disposal. Nevertheless, within a few hours the girls had brought most of the Yarmouth yards to a standstill and the strike had spread to Lowestoft. The curers immediately stopped buying herrings and fishermen were unable to sell their catches. A few herring were sold to the freshing trade and to the smokehouses, but the girls had as good as paralysed the industry. Within two days the curers capitulated and the women's barrel money was restored to 1 shilling.

HERRING PEOPLE: Isobella Eunson

When Isobella joined the industry in the mid-1950s, she became one of the last generation of herring girls. She came from the island of Whalsay in the Shetlands and had already worked two Shetland seasons when in 1956, aged 17, she and two friends decided to go as a crew to Yarmouth, working for Bloomfield's. By the second half of the twentieth century, some things in the herring industry had changed – they had suitcases rather than kists and they travelled by scheduled trains instead of specials – but some things were just the same: outdoor work, cloots on your fingers, cuts and sores, three to a bedroom.

We didn't go herring gutting because we particularly wanted that work or for the money. It was for the adventure, the fun and excitement. We stayed in lodgings with Mr and Mrs Cumby in Exmouth Road, which was very close to the Fish Quay and the curing yards. Bloomfield's paid for our room but we had to buy our own food, which Mrs Cumby cooked for us. From our lodgings it was only a short walk to work.

Once the herring arrived at the yard we would keep working until they had all been gutted and packed, no matter how late in the day that might be. I think the latest I ever finished was ten o'clock at night. If there were no fresh herring being landed or if it was our day off, we would go and pack kippers for Mac Fisheries in order to earn a little spending money. Some of the girls even got casual work in a factory, canning potatoes. That year I think we were the only crew from Shetland but we got to know other crews who came from Portknockie, Peterhead, Buckie and Fraserburgh and there were also some girls from Ireland.

We worked hard during the day and in the evenings we went round the attractions in Yarmouth. When I first saw Yarmouth I was amazed at how big a

Herring girls on strike. This is 1936 and the herring girls went on strike demanding that their barrel money be restored to one shilling per packed barrel. For two days the fishing at both Yarmouth and Lowestoft was at a standstill.

place it was and how many people there were. There was so much to do and lots of places to go for entertainment and it was just so different from home. Our days were so full that we never really had time to feel homesick and all the Yarmouth people that we met were very nice and friendly towards us.

When we returned home at the end of the season we had no money left. We had spent it before we left Yarmouth, mostly on new clothes and make-up.

Timelines Chapter 7

1899 Miss Davidson's Rest House for Scottish Fisher-Folk opens in Yarmouth

1905 Corridor coaches with toilets are used for the first time on the herring worker trains

1911 7,000 Scots girls travel south to work in East Anglia

1911 It is reported that the Scots girls on average take home £16 from the Yarmouth season

1912 Statistics record that 10,800 Scots women depend on the herring industry for their living

1913 Mrs Stewart of London funds a shelter at Yarmouth for the herring women

1924 One crew of herring girls gutted and packed thirty-seven barrels in one day

1925 Thirty special trains bring herring workers from Scotland to East Anglia

1929 The Red Cross opens a Dressing Station at Great Yarmouth for the herring season

1935 The Scottish curers cut the money paid to herring girls by 2d a barrel

1936 Scots girls strike in Yarmouth and Lowestoft for 1s per barrel

1938 Thirteen special trains bring herring workers from Scotland to East Anglia

1938 Scots girls strike in Yarmouth over Sunday-caught herring and living allowances

1949 The Scots girls go on strike in Yarmouth for higher ration allowance

1953 The Church of Scotland opens a new canteen in Lowestoft for the fish workers

1960 Herring Industry Board sanctions just three crews to work the Yarmouth season

By the 1950s the age-mix of the herring girls was changing as fewer young women entered the industry.

CHAPTER 8

Herring Ports and Stations

There were a great many fishing ports and small harbours around the United Kingdom coast but not all of them became recognised as herring stations. A herring station was defined as a harbour which received landings of herring where the herring were processed for export. The term 'station' could therefore be used to describe a large port like Grimsby or a small harbour like Hopeman; it could sometimes refer to a windswept bay on a remote Scottish island. By the twentieth century some of the country's largest fishing ports, such as Hull, Fleetwood, Grimsby and Aberdeen, were primarily handling trawled white fish, but during the course of a season they would also receive landings of herring. If herrings were landed at one of these ports but there were no curers or Klondykers present (the fish all being sold fresh or for kippering) then, technically, that port did not fit the definition of a herring station.

At the lower end of the scale, there were numerous small harbours and fishing settlements where regular (but small) landings of herring were made, even though at some, such as Cromer and Dungeness, the fish had to be landed on the beach. In these small locations the herrings were caught by small inshore vessels and were sold to the local home market. The landings, however small, were all counted, officially recorded and included in the total British catch.

In between the large ports and the tiny ones was a tier of herring ports whose good harbours and proximity to herring stocks meant that they frequently attracted fishermen from other parts of the country and, often, the Scotch curers. Like the major centres such as Wick, Fraserburgh, Yarmouth and Lowestoft, these ports were regular fixtures on the calendar of the British herring industry. By 1900, the use of telegrams and telephones had speeded up communications within the industry and helped weld it into a single national enterprise rather than a series of separate regional fisheries. If an East Anglian skipper was about to head to Northumberland but heard that the fishing was better at the Isle of Man, he would sail there instead, and maybe he would be joined there by Moray Firth vessels who were working their way from the Shetlands to southern Ireland.

For a variety of reasons (disappearance of shoals, changes within the industry and so on), landings at all stations and ports could vary in size from year to year and decade to decade – this is illustrated in Appendix 6 (English landings) and Appendix 7 (Scottish landings). The following random selection of ports has been picked to illustrate that there was more to the industry than East Anglia and the Moray Firth and that, during the first half of the twentieth century, smaller ports played their part in the national herring industry.

Plymouth

Plymouth is a port that we associate more with the Royal Navy than with fishing, but during the winter months herrings were caught and landed there. As winter was a quiet time on the east coast, it became common practise at the end of the nineteenth century for East Anglian drifters to sail to the south-west for the winter months. In the early twentieth century, Lowestoft steam drifters were a common winter sight at Plymouth, working alongside smaller West Country sail and motor vessels. The catches off Plymouth were tiny in comparison to those in the North Sea but they were sufficient to keep the boats employed and owners were happy with the additional cash flow, which was better than laying the boats up for the winter. The Lowestoft owners showed little seasonal generosity to their crews because they expected them to work through the festive season – in 1935 it was reported that the east coast fleet had fished on Christmas Eve night and landed 130 crans of herring at Plymouth on Christmas Day morning.

The herring grounds were about ten miles south of the Eddystone lighthouse, well out into international waters. In 1913, when landings in East Anglia reached their peak, landings at Plymouth were a modest 23,000 cwts but during the 1920s they grew to over 100,000 cwts a year. However, a combined fleet of around 150 East Anglian steam drifters plus a similar number of local boats and some thirty to forty French steam trawlers put intense pressure on the herring population. Not surprisingly, in the 1930s the herring stock collapsed, never to fully recover – in 1938 a mere 560 cwts were landed.

Newlyn

Like other Cornish ports, Newlyn was known more for its pilchard fishery than for herrings but during the winter months, along with near neighbours St Ives and Padstow, it often saw quite reasonable herring landings. In 1928 nearly 8,000 cwts of herring were landed in Newlyn and 4,000 cwts in 1938. Because there was a railhead just over a mile away in Penzance, it was relatively easy to convey fresh fish from Newlyn to the rest of the country. Fishermen from Newlyn and other Cornish harbours often joined herring fisheries elsewhere around the country, such as those at the Isle of Man, Yorkshire and East Anglia.

This commemorative plaque at St Ives harbour, Cornwall, is a reminder of how the herring industry linked together different parts of the country.

The small volume of herring landings at Newlyn was not large enough to attract herring curers and exporters. However, during late winter and early spring the port hosted large numbers of Lowestoft steam drifters who were landing mackerel rather than herring. Like herring, mackerel are a pelagic fish and can be caught by using drift nets with a slightly larger mesh size.

Since the 1880s, the Lowestoft drifter fleet had become a familiar sight at Newlyn when they arrived after the winter herring season at Plymouth. The presence of a large, modern, well equipped fishing fleet on their fishing grounds was something that the Cornish fishermen with their much smaller boats found hard to accept, especially as, contrary to Cornish custom, the East Anglian boats fished on Sunday. In May 1896 simmering resentment among the Cornishmen culminated in the Newlyn Riots, two days of mob violence and attacks on Lowestoft vessels. The rioting was only quelled when the government sent in a contingent of 300 armed troops and three Royal Navy vessels. In the fraternity of the sea these events were soon forgiven and forgotten (although some months afterwards, a Cornish boat that arrived in Lowestoft for the East Anglian herring season was stoned by local youths).

In the late nineteenth and early twentieth century, Newlyn attracted artists such as Stanhope Forbes and Walter Langley, founders of the Newlyn School of artists, who produced paintings depicting local scenes and local people. Newlyn's residents, especially the fishermen, could earn extra money by posing for the artists. In the twenty-first century Newlyn is still one of the busiest trawling ports in the United Kingdom but there are few opportunities for spare-time modelling.

Peel

The Isle of Man was historically the centre for herring fishing in the Irish Sea. For much of the nineteenth century, Cornish and Scots fishermen had come there in pursuit of its autumn-spawning herring. Early in the year, Manx fishermen used to sail north to join the Scottish herring fishing before returning to the Irish Sea for their own local herring season, which peaked in August and September.

By the early twentieth century the Manx herring industry was based on the south-west of the island, at the harbours of Peel, Port Erin and Port St Mary. Port Erin was home to a small fishery research establishment run by Liverpool University but Peel was the busier herring port. The most famous food product on the Isle of Man is the Manx Kipper and many of the herrings landed there were destined for the local smokehouses. However, the Scotch curers were regular visitors and they bought herrings which were salted and barrelled for export. In 1930, five curing firms worked the Isle of Man season, employing twenty-four crews of Scots

STEAM DRIFTERS AND FISHING BOATS IN PEEL BAY I. O. M.

This photograph from around 1912 was taken at Peel harbour, Isle of Man. The steam drifter in the centre of the picture is the *Majestic* INS 151 from the Moray Firth village of Hopeman.

girls. In 1960, thirteen crews and two gutting machines produced a total of 2,187 barrels of salted herring.

In 1913 herring landings at Peel were over 110,000 cwts but by 1928, despite there still being some seventy drifters fishing out of the harbour, landings had dropped to 32,000 cwts. The catches remained steady through the 1930s; in 1938, when 28,000 cwts were landed, the Scots curers produced 3,700 barrels for export. Well into the 1950s, Bloomfield's of Yarmouth were still sending boats and agents to Peel for the herring season. In 1958, they presented Peel police station with one of its great unsolved cases when an intruder wandered into their curing yard office and stole £9 10s from the petty cash box. The loss had to be accounted for in great detail to the insurance department at Unilever's head office.

The Isle of Man herring fishery, like so many others, was destined to collapse through over-fishing, although this did not happen till the late 1970s.

Tarbert

At the northern end of the Kintyre peninsula is the small Argyllshire port of Tarbert. Situated where the long finger of Loch Fyne joins the Firth of Clyde, Tarbert was an ideal base from which to pursue the Loch Fyne herring. This herring stock made their annual spawning migration from Loch Fyne to the Maidens Bank, off the south Ayrshire coast. Tarbert fishermen were at the forefront of developing and adopting the ring-net (see Chapter 5) for herring fishing and they frequently spent long periods away from home, fishing the Minches, the Irish Sea and, in the 1950s, reaching as far as Whitby on the Yorkshire coast. Vessels (sail initially, then motor) fished in pairs, surrounding the herring with a long net and then scooping them out of the sea on to the boats. Steam drifters were rare visitors to this area.

Tarbert had an autumn and winter herring fishery that yielded steady if unspectacular catches, 3,500 cwts in 1928 and 10,000 cwts in 1938. Around half of these catches were salted by curers for export; of the rest, some ended up as Loch Fyne Kippers and some were sold to Klondykers. Tarbert had no rail link so most of the fish landed there were transported away by sea. There was a rise in catches in the late 1940s – in 1949, 280,000 hundredweight were landed and 2,000 barrels of salted herring produced. However, catches dipped away again and when 28,000 hundredweight were landed in 1955, a mere seventy-six barrels (250 hundredweight) of salted herring were produced.

There was one sales outlet which was unique to Tarbert and the surrounding area. Because it was only a short distance by sea to Glasgow, local fish wholesalers developed a regular trade through which this large city was provided with daily deliveries of almost sea-fresh herring. Small steamers would rendezvous with the fishermen at the fishing grounds to buy their catches and then transfer the herrings from fishing boat to the carrier. Sometimes they would take as many as 1,000 boxes of herring at a time to Glasgow. By this means, it was often possible to buy herrings in Glasgow which only five hours earlier had been swimming in the sea.

Stronsay

Steam drifters had proved that they could quickly get to port with a catch of herrings but sometimes the return distance from herring ground to a permanently established port was just too far to comfortably travel daily. Therefore, in more remote areas it was necessary to establish temporary herring stations close to the herring fishing grounds.

The island of Stronsay was known to have been used in the Stewart era as a temporary fishing base by Dutch herring fishermen. In the 1920s and 1930s it became a major player in the British herring industry and the biggest herring port in the Orkneys. During this period, the island's resident population of about 1,000 was joined during the summer months by over 4,000 migrant seasonal herring workers: fishermen, gutters, merchants, curers and coopers. A fleet of some 300 steam drifters from around the British Isles fished the herring grounds to the east of Stronsay and then hurried them to the sheltered harbour at Whitehall Village, where the shore-based workers were kept busy processing the catch. There was so

In this view of Stronsay, Orkney, from before the First World War, the harbour contains a mixture of sailing drifters and steam drifters from Moray Firth ports as well as visiting cargo vessels. On the shoreline two young boys (future fishermen, maybe?) play with a toy yacht.

much congestion around the harbour that overflow curing yards were established on the small offshore island of Papa Stronsay.

In 1928 herring landings at Stronsay were 193,000 cwts (about half the figure for Wick, which was on the mainland but had the benefit of a rail link). At Stronsay that year 109 coopers and 550 gutters and packers produced 70,000 barrels of cured herring for export. The figure of 550 gutters is only a third of the number who that year worked at the large fishing port of Peterhead but, considering the remoteness of Stronsay, it is quite impressive. The remainder of the Stronsay catch was mostly sold to Klondykers, so there was a steady stream of cargo ships in and out of the harbour, loading fresh herrings and barrels of cured herrings to take to Germany and the Baltic countries. Incoming cargoes included coal, salt and new barrels. The large British herring companies were regularly represented at Stronsay, including Bloomfield's and their curing subsidiaries, Joseph Slater and Wm Low. During the 1930s, catches started to fall as the herring stock began to diminish – in 1938 they fell to 54,000 cwts. That summer Stronsay hosted drifters from Buckie, Inverness, Wick, Fraserburgh, Peterhead, Yarmouth and Lowestoft and the forty-two coopers and 129 gutters present on the island produced 20,000 barrels for export. In late 1938 Bloomfield's received a string of complaints from the Baltic States (see Chapter 6) about the Stronsay cure and they found that the problem had partly been the standard of the gutting. When fishing resumed after the Second World War, the herring had disappeared from Orkney waters and Stronsay's heyday was over.

Anstruther

The harbours of Anstruther and Cellardyke are located close to each other on the Fife coast. Although herring landings at these small towns were recorded jointly in official statistics, Anstruther was the main player in this partnership, having a larger harbour than its neighbour and providing better facilities. The Firth of Forth was home to the longest-established commercial herring fishery in Scotland and for centuries the villages on the Fife coast participated in it. In the early nineteenth century fishermen and curers from this area began to travel north for the summer fishing at Wick and the Shetlands. By the late 1860s, some were also making their way to East Anglia for the autumn fishing.

The main herring fishery in the Firth of Forth was during the winter; there was another appearance of herrings in late summer, known locally as the 'Lammas Drave', but this faded after the First World War. Another type of fishing practised on this coast was 'great line' fishing, which involved laying out long lines to which were attached hundreds of baited hooks in order to catch demersal species such as cod. This type of fishing had a close relationship with herring fishing because the line fishermen bought herrings to use as bait. If there were no herring being caught, the liners had a struggle to find alternative bait.

Anstruther harbour is pictured here during the herring season with steam drifter and quayside activity at its height.

Fife boat owners were quick to join the rush for steam, accounting for the large numbers of 'KY' (Kirkcaldy) registered steam drifters seen in herring ports from Shetland to East Anglia. In the autumn of 1911, Anstruther and Cellardyke supplied thirty-two of the 549 Scots drifters working at Yarmouth, and in 1929 there were thirty-five locally owned steam drifters. Between herring seasons the Anstruther/Cellardyke steam drifters were sometimes used for great line fishing. The last steam drifter to be built in the United Kingdom, *Wilson Line*, was delivered to a local owner in 1932.

By the twentieth century, the Ansthruther herring fishery was diminishing in importance but the total herring landings for 1929 (predominantly from the winter fishing) were a creditable 61,000 cwts. The number of barrels of cured herring produced that year in the whole district was just 356, the greater part of that year's landings being sold fresh (Anstruther was served by a railway line) or Klondyked. At the height of the season (between February and March), local boats were joined by ring-netters from the Firth of Clyde and steam drifters from the Peterhead, Buckie and Inverness districts.

By 1949 the total landings of herring at the two harbours had dropped to 180 cwt and it was no longer worthwhile for the curers to put in an appearance. In

1950 there were still twenty-four steam vessels registered at the twin harbours but they mainly fished away from home and landed their catches in other ports (Aberdeen, Stornoway, Lerwick, Shields, Yarmouth and Lowestoft).

In 1969 Anstruther became the home of the newly-established Scottish Fisheries Museum, a fascinating place if you are interested in sea fishing and well worth a visit even if you are not.

HERRING PEOPLE: Peter F. Anson

Peter Anson was a deeply spiritual man who initially studied architecture, then trained as a Roman Catholic priest, but became a well known chronicler of the herring industry. He travelled widely, writing on religious matters, making drawings of churches and spending several years in various Catholic monasteries.

In the 1920s and 1930s, between travels, he lived for some time in the Moray Firth fishing towns of Portsoy, Banff and Macduff. Here he developed a deep interest in the lives of the herring fishermen, formed many friendships among them and developed an encyclopaedic knowledge of the fishing industry.

Over his life he wrote many books, including several about fishing and fishermen. He recorded the herring industry while it was still vibrant and economically important to many people's lives. He produced drawings and many watercolour paintings (at least 800) of boats, herring fishermen and their villages. When it was planned to set up a fishing museum for Scotland, he was an obvious first choice as curator, a post he held until the museum was up and running.

North Shields

The Tyne is a deep harbour with access at all tides and North Shields was a busy year-round trawling port which, during the summer months, became a major herring station. As the port's hinterland included many large towns and industrial conurbations, there was a good local demand for fresh fish. It was at North Shields that Richard Irvin had built up his fishing empire, which included catching, selling and canning herrings.

In the twentieth century North Shields regularly recorded the highest herring landings of any English port after Yarmouth and Lowestoft. They rose from 300,000 cwt in 1908 to 500,000 in 1913 and then dropped steadily to 134,000 by 1955. The number of steam drifters registered in the port peaked at twenty in 1908 – most herring landings were made by drifters visiting from Scotland and East Anglia. Herring were caught from May to October, but the largest landings were during July, August and September, the herring being of a separate stock to those found off East Anglia.

This advert in
Fishing News lists
all the services
and facilities
provided at
North Shields
and tells us what
fishermen needed
in a harbour (ice,
water, coal and

THE FISHING NEWS.   Friday, August 15, 1913.

NORTH SHIELDS HERRING MARKET.

This market has now fallen into line with other herring ports, the old system of selling and discharging by means of bogeys and baskets having been abandoned, and the cran measure, with buyers accepting delivery at the boatside, adopted.

EVERY POSSIBLE CONVENIENCE.   ADDITIONAL LANDING SPACE.   RAPID DISCHARGING ASSURED. NEW STANDS FOR PACKERS AND PICKLERS.   CHEAP WATER AND COALS.

Abundant supplies of Ice, Fish Boxes, etc.   Provision for Drying and Storing Nets, etc.   All the Leading Buyers and Inland Markets represented at the Port. Smoke-houses, Canning Factories, etc. Adequate means for quick dispatch by sea or rail.

Applications for Packing Stands, etc., and for general information should be made to the Quaymaster (Mr. THOS. McKENZIE), North Shields.

Scots herring girls were a common sight at North Shields and their contribution to the industry and to the local economy is commemorated in a wrought iron artwork set in the wall of the Fish Quay. In the early 1950s, Vernon Green, son of 'Wee' Green, worked for the fish sales company of Norford Suffling at their office on the North Shields Fish Quay.

Whitby

Whitby, a centuries old fishing port, famous for its connections with Captain Cook and Count Dracula, was something of an anomaly in the herring industry of the early twentieth century. From 1900 onwards, while other ports experienced steadily increasing landings of herring, those at Whitby declined. In 1900, Whitby had been a large herring fishing port and during the Yorkshire herring season (August to September) it attracted sailing drifters from Scotland and East Anglia as well as luggers from Cornwall. As at Yarmouth, herrings were still sold by the 'last' and in Whitby the counting was done by old women and widows who were paid 1s 6d a day to manually count the herrings. By 1905 the Cornishmen had stopped coming, and by 1913 no significant landings of herring were made at Whitby. Herrings were still caught off the Yorkshire coast but they were being landed at Scarborough and at Hull. Some landings were made in the period between the wars but generally they were poor – in 1928 there were two steam drifters registered at Whitby but herring landings for the year were a mere 500 cwts.

Whitby re-emerged as a herring port after the Second World War. The herring stocks had recovered during the six years when very little fishing had been permitted off that part of the coast. Scots fishermen returned to Whitby in the late 1940s, some now using ring-nets instead of driftnets. By 1948 well over a hundred vessels were landing increasing volumes of herring at the harbour, once again making Whitby the exception to other herring ports, which were now seeing declining catches. The Whitby herring landings regularly attracted Scots curers and Scots herring girls were a familiar sight on the quayside. In 1948 herring landings were 106,000 cwts (compared to just 72 cwts in 1938) but by 1955 they had eased to 64,000 cwts.

This impressive building is next to the Fish Quay in North Shields and was for many years the headquarters of Richard Irvin, one of the biggest fish selling companies in the country. It has now been converted into apartments.

The focus of this 1900 Whitby photograph is the fine schooner, probably involved in herring exports. The small fishing boats to the left of it have PZ (Penzance) markings. Probably from Newlyn, they have come to join the Yorkshire herring fishery.

CHAPTER 9

1939 to 1960

The Second World War

In September 1939 war was declared and, just as 25 years previously, large numbers of steam and diesel fishing boats were requisitioned by the Admiralty. Fewer drifters were requisitioned than in the 1914–18 war but the numbers included some 300 that had seen service during the First World War. As in 1914–18, the drifters were allocated to routine naval support duties around the coasts. The flotilla of vessels sent to the Dunkirk evacuation included forty-one drifters, one of which managed to ferry back 236 British soldiers in one night. Of the seventy-three steam drifters lost between 1939 and 1945, six were lost at Dunkirk.

Some of the motor fishing vessels operated by Scottish fishermen were deemed too small for the Navy's requirements, and so were allowed to carry on fishing. As in the First World War, large areas of the North Sea were mined, so most fishing was on the western side of the British Isles. In 1939, in the very early weeks of the war, the Admiralty allowed a limited autumn herring fishery to take place off East Anglia. The small fleet engaged on this fishery was monitored and protected by air cover from RAF Bircham Newton in Norfolk, in an operation that the airmen referred to as the 'Kipper Patrol'. This fishery was not repeated in subsequent years but, later in the war, a small summer herring fishery was permitted off the Northumberland coast. This was based at Seahouses and it brought Scottish ring-netters for the first time to the east coast of England. Although good numbers of herrings were caught, the fishery was not a great success because there were difficulties in finding enough lorries to transport the fish to the wholesale markets.

Not surprisingly, English herring landings plummeted during the war years. In the safer Scottish waters the herring catches of 1940 and 1941 were about half of pre-war levels, but in the following years they rose to around seventy per cent of the 1939 catch. Fish was not rationed but the government operated a control price of £4 18s a cran for fishermen landing herrings.

In this war some of the Bloomfield's drifter/trawlers remained on civilian fishing duty. As well as operating this small fleet, Bloomfield's were given the wartime responsibility of managing a fleet of twenty Dutch steam trawlers and their crews. These vessels had been away at sea in 1940 when the Netherlands was rapidly overrun by Germany and they had been ordered by their own government to head for English ports rather than return home. As a result, the Dutch fishermen spent five years of the war in exile living in and fishing out of Fleetwood.

HERRING PEOPLE: Tom Bruce

For Tom Bruce, this war appeared to be going a bit better than the previous one. He was now in his fifties and his boat, the *Prevail*, was too small to be requisitioned by the Admiralty. With a crew of older men and young boys, he was able to carry on fishing, mainly off the west coast of Scotland.

In February 1944 the *Prevail* was part of a group of boats that were fishing out of Ayr when bad weather hit during the night. Some boats stayed out at sea, but Tom thought it would be wisest to seek shelter. While making for harbour under wartime black-out restrictions, the *Prevail* was run down in the darkness by a Belfast-bound collier and Tom and three of his crewmen were lost. There were only two survivors.

The Industry Picks up Again after the War

Once the war was over, commandeered fishing boats were returned to fishing duties. Herring fishing resumed fully in 1946, following the traditional cycle of the Scotch Voyage followed by the East Anglian season. Soon, demand from the Continent picked up again and the curers got down to business, providing work for coopers and herring gutters. The gutters were certainly busy – John Oakley of Norfolk recalls that on leaving school in 1948, one of his first jobs on the farm at Witton was to empty a whole lorry load of herring guts from Yarmouth and to mix it with chaff and straw before spreading it on the fields as manure (not a particularly fragrant task).

The Herring Industry Board soon resumed its efforts to prevent gluts in the market by monitoring catches and controlling fishing. One of its methods was to impose a limit on the number of nets that drifters could carry; sometimes it was as low as sixty-three (most steam drifters were capable of working ninety nets). Gerald Tungate, a Younker on the Winterton steam drifter *Romany Rose* (YH 63), recalls that on the Scotch Voyage in 1946 all boats were limited to seventy nets. Despite this restriction, on their first night's fishing at Fraserburgh they landed 120 crans – a very good haul. This prompted the Board's representatives to order the *Romany Rose* to stay in harbour for the next three nights, much to the disgust of her skipper, 'Toody' Rudd. To fill the time usefully, Toody had his crew tan and dry a new set of nets. At some ports the Board introduced 'pooling' arrangements

This building is one of very few former beating chambers in East Anglia that have-not been demolished or converted to domestic use. It was owned by Toody Rudd, skipper of the *Romany Rose*.

under which boats arriving within a certain time band (say, between 8.00 a.m. and 10.30 a.m.) would receive a pooled, or average, price for their landings. A second pool later in the day might produce a lower average price.

Most English herring were still being caught with driftnets but there were signs of change. The Scots ring-netters that had been a feature of the pre-war herring fishery in west Scotland were now increasingly seen in England. The Scandinavians had begun to design mid-water trawls with which herrings could be caught during daylight hours, before they rose to feed near the surface. British boat owners watched this development with interest and in the early 1950s Bloomfield's carried out several trials using new Dutch and Swedish trawls and reported 'there is little doubt that the Larsen trawl is deadly given a reasonable show of herrings on the echo meter'.

An important event for the crews of fishing boats was the introduction in 1948 of a national minimum wage for share fishermen. This meant that there would be no repeat of the poverty that Scottish fishing communities in particular had experienced during the 1920s and 1930s. It was probably not a popular measure among the boat owners as the minimum wage came from their pockets; some English companies had already introduced a similar scheme back in 1936.

Nevertheless, Scots fishermen were still leaving the herring industry because they felt there was a better living to be made from trawling.

The herring industry was still a large employer and there was confidence that it could continue to thrive. As the 1950s arrived, companies such as Smalls of Lowestoft and Bloomfield's of Yarmouth were still building diesel drifter/trawlers, and fish companies such as Richard Irvin and Suttons were still prospering at Yarmouth and at other herring ports.

HERRING PEOPLE: Gerald Tungate

In 1946, the Prunier Trophy was re-introduced after its wartime absence. That season there were some 300 boats fishing out of Yarmouth and Lowestoft. One of these was the Winterton steam drifter *Romany Rose* (YH 63), skippered by Walter 'Toody' Rudd. Gerald Tungate was one of his youngest crewmen.

Gerald relates:

Up to November we had a very quiet Home Fishing. On the Sunday (3 November), we left Yarmouth and steamed out for about three hours to beyond Smiths Knoll. I was Younker then and when I was taking a turn in the wheelhouse, I had to note any signs that herring might be around – things like seabirds, whales, porpoises and the colour of the water – and to report them to Toody. At about three-thirty he headed us to the spot he had decided on and he positioned the *Romany Rose* so that the current would carry us into an area where I had earlier seen whales. With dusk falling, we shot the nets (we were all limited to seventy nets that year). We were using a new fleet of nets that were not fully softened by the seawater so they were still a bit stiff to handle and hard on your fingers.

The sea was calm and the weather was good. At six o' clock, the Skipper ordered a look on so we lifted the first net but there was hardly any herring on it so we slacked it off again. By midnight there was a lot of whale activity around the boat (we couldn't see them in the darkness but we could hear them blowing) and we took another look on. This time we got a 40 cran shimmer and so we began to haul. The first thirty or so nets had average herring numbers but once we got to the half-way bowl (the buff attached to the middle net) we were getting a 100 cran shimmer. Hauling slowed down because everything became much heavier and the real hard work began. We soon found nets that had fish caught on both sides of them, so we knew there had been a double swim. The quantity of herrings made the nets really heavy and it was difficult to shake the herring off the nets – we had to shake the net once to release the fish on one side and then flip it over and shake it again – and all the time those new nets were cutting into our hands! We then came to some nets where the buffs had disappeared, pulled beneath the surface by the weight of the herring in the net, and these were even heavier to lift on board. There were some nets where it seemed that there wasn't a single mesh that didn't have a herring caught in it! We had no time for a break but every so often the Mate, Johnny Goffin, brought us all a cup of tea, which we drank standing out on the deck.

The fine, calm weather had made our work easier in the beginning but now we could have done with a bit of a swell because, when the boat rolls, the motion tips the herrings down the scuttles for you and the rocking helps to spread them evenly around the hold. We had never worked so hard in our lives and by late morning we had been at it for twelve hours but we knew now that we were on for at least 200 crans. We had herrings packed everywhere – even the tan tank in the bow had been emptied and was full of herring.

Sometime after mid-day we emptied the last net and Toody began to head very gingerly for port. Because we had herrings in the tan tank her nose was low in the water and so the sea was running in through the hawse pipe and washing across the deck. Toody had radioed ahead to Yarmouth and when we berthed late in the afternoon a relief crew was waiting to take over from us and to unload the herring. By then we were all practically dead on our feet. Toody told us to go home, get a good night's sleep and be back on the boat at eight in the morning.

The catch sold for £950 and the following day it was officially confirmed as 246¾ crans, the highest of the season so far! We then had to wait till 22 November, when the competition officially closed, before we knew for sure that we had won the Prunier Trophy – the first time that it had been won by a Yarmouth boat!

HERRING PEOPLE: Jack Stowers

By the late 1940s the herring industry was back in swing, and so was the railway service transporting the fish from the coast to the major cities. Jack Stowers, a locomotive fireman in those post-war years at Yarmouth, tells what it was like to be working these trains:

During the herring season there was a daily herring train from Yarmouth Vauxhall to the north of England. It was a favourite turn for us engine crews because you were sure to get some overtime. We signed on at the engine shed at 12.45 and while we were getting the loco ready, the goods yard would be a bustle of activity as lorries of boxed herring arrived from the fish market to be loaded into the goods vans. There was a different van for each town; Manchester, Sheffield, Leeds and so on. They were ordinary unrefrigerated vans; at other times of the year they were used for carrying fruit from Wisbech and bags of sugar beet pulp. When they were returned empty to Yarmouth they had to be scrubbed out as the fish boxes always leaked.

At 2.30 we set off with around twenty vans to Reedham Junction, eight miles away, where another twenty vans that had been brought up from Lowestoft were coupled on to our train. After about thirty minutes we pulled away again. The train was classified as an express freight (though you wouldn't know it sometimes!). After Norwich we went via Thetford, Brandon and Ely. Our highest speed was around fifty miles an hour but because of frequent signal stops and heavy rail traffic, we rarely reached that.

The fish trains were always smelly but because we were on the locomotive at the front of the train we never noticed it – the guard, in his brake van at the rear, used to get it all! The melting ice, mixed with the oil from the herrings, leaked out through the floor of the vans and on to the track. This could make the rails very greasy for the trains following behind us.

Sometimes it could be seven in the evening before we reached the exchange sidings at March. Here we came off the train and another loco took it north towards Doncaster. We topped up the coal and water, turned the loco on the turntable, took a meal break and waited to be allocated an eastbound freight, usually a coal train, to haul back to Norwich. It could be three in the morning before we signed off at Yarmouth.

In this image from the 1950s the locomotive of the London fish train is getting up steam as the last few boxes of herring are lifted from lorries into rail trucks at Lowestoft goods yard.

At the engine shed we used to run our own version of the 'Prunier Trophy' – at the end of the herring season we used to tot up who had clocked up the highest number of overtime hours during those ten weeks!

The Herring Industry Board

The Board continued to work at making the industry profitable. In 1947 it opened its first herring reduction factory – the Norwegians and Danes had been operating them for years. These factories processed whole herrings into animal feed and extracted the fish oil, which was used in the manufacture of ice cream and margarine. When the freshers, smokers, curers and canners had bought all they required, the fishermen could still make a sale, albeit at a lower price, to the reduction factories. These factories became increasingly important to the British herring fishermen. In 1947, the year the first factory opened, 3 per cent of the total British herring catch was processed this way, but by 1953 45 per cent of the national catch was going to the reduction factories. Sadly, given that the original aim of the competition had been to promote home consumption of fresh herrings, some of the Prunier winning catches now ended up being sold to the Yarmouth reduction factory.

The Board continued its advertising campaign to promote sales of herring in the United Kingdom. This included holding 'herring weeks' around the country and providing promotional leaflets to retailers and educational material to schools.

In the summer of 1950, as herring prices began to fall, the Board found itself caught between the austerity-minded government and the angry Scots fishermen who were demanding support in the form of minimum prices. The price paid for herrings by the reduction factories was already subsidised so the government refused to offer any further help.

The Board funded research in a variety of areas, including the quick freezing of herrings, new methods of catching fish and how to prevent the tomato puree in tinned herrings from turning brown during the canning process. From the mid-1950s, the Board became particularly interested in ensuring that the consumer received good quality products and organised inspections of smokehouses to check the quality of their product. By 1954, the Board was putting pressure on smokehouses to abandon the practice of artificially colouring kippers and so the 'Painted Ladies' (see Chapter 4) were gradually replaced by better quality kippers that had a natural golden colour.

In the late 1930s some drifter owners had begun to pack herrings with ice into wooden boxes in the hold of the drifter, instead of tipping them in loose and then shovelling them out on landing. The Board now decided to encourage this practice and made loans available to fishermen for buying aluminium boxes in which to pack their herrings.

This image of the Scots drifters moored up at the weekend in Yarmouth was taken in the 1950s; by then there were very few steam drifters among them.

By 1958 the Board owned reduction factories at Yarmouth, Fraserburgh, Stornoway, Wick and Peel and there were privately-owned factories operating at Lerwick, Peterhead and Ardglass (Northern Ireland). The meal and oil branch of the trade, which had done well in the early 1950s, started to fade at the end of the decade when cheaper alternatives became available from Peru. This, together with falling herring catches, led to the moth-balling of some of the British reduction factories.

HERRING PEOPLE: Sandy McPherson

By 1950 there were no steam drifters remaining at Hopeman but there was one motor drifter that still travelled around the British Isles in pursuit of herring. Sandy McPherson (tee name Sandy 'Daich') joined the *Fair Morn* (INS 195) in 1950 as a deckhand learner and later became cook. That autumn he went to Yarmouth for the first time and he recalls his experiences:

> We left Yarmouth each afternoon and shot our nets in the early evening. We mostly fished round Smiths Knoll but we did once go down to fish on the Sandettie Bank off Dunkirk.

Sandy McPherson at the age of 16 while working as cook on the Hopeman motor drifter *Fair Morn*.

During net hauling my job was to coil the leader rope into the 'box', which was a small hold in the bows of the boat that you climbed into through a small hatch on the deck – it was not pleasant being in there. The rope was about three inches thick and was coated with tar. It was fed down to me and I coiled it flat on the floor from the walls of the box into the centre. It was dirty and difficult to manipulate – I had to stamp on it to make it lie flat. When one layer was complete I started the next and so on until all one and a quarter miles of it was neatly stowed away. The box would be nearly full by then. Hauling the nets could take four to five hours for a decent catch and about three hours if it was a poor haul.

The rope was a vital piece of equipment and if any part of it got damaged it would need urgent repair. I had to leave any frayed section as a loop out on the deck and then continue to stow away the rest. Once hauling had finished, we filled a bucket with hot water from the kettle and soaked the loop in it. This softened the tar so that we could make a splice to do the repair.

When we got to the quayside the mate jumped ashore with a sample basket of herring and took it to the sale ring to be inspected for quality. Once the catch was auctioned, porters would arrive with swills (baskets) and lorries and we unloaded the catch. The *Fair Morn* had an arrangement with Richard Irvin Ltd, the fish sales firm, who had offices at the Fish Quay. They would sell our catch for us and then settle up with the skipper at the end of the season.

The motor drifter *Fair Morn* at the mouth of Yarmouth harbour in the early 1950s. She was a typical Scottish motor drifter and was the last drifter to work out of Hopeman, Moray.

As soon as we arrived at the fish quay, 'runners' from the butchers, bakers and grocers would jump on board and take our orders. They delivered these before we sailed again and they left the bill with Richard Irvin Ltd for payment. Most food was still on ration in those days and, although fishermen were entitled to extra coupons, there was not a lot of variety, especially regarding meat. We never bothered with fresh milk but used 'Carnation' tinned milk instead.

There were no regular mealtimes on board, except maybe while we were heading out to the fishing grounds. When I was cook I used to make a big suet dumpling two to three times a week or sometimes a thick soup. On Sundays we had a stew or very occasionally roast beef. Herrings were a staple food on board and the crewmen were quite fussy about their herring and liked to select their own, usually the ones with plenty of roe. When I cooked the herring I used to fry them (there was a lot of fried food on board!). We occasionally had a pudding, maybe rice boiled in milk, or a bread pudding but more often it was just 'Typhoo Pudding' – a cup of tea.

On Saturdays, after unloading the catch, we moored up in Great Yarmouth as close as we could to the Town Hall. One of the first places I called at was the rock shop down Regent Road, where I used to swap some of my sugar ration for sticks of rock. I then joined the queue of Scots fishermen at the Post Office, opposite the Town Hall, to send my dirty clothes home in a canvas 'baggie'. In the middle of the dirty washing I would have stuffed a few sticks of rock for the folks back home.

Shoe-shine boys used to come on board the boats to see if the men wanted their shoes polished before they went out for a night on the town. Some of the crewmen went to the cinema and then had a pint in a pub. Each Saturday I received ten shillings (50 pence) spending money – the rest of my pay came at the end of the trip. We young boys used to wander around the town and along the seafront. We liked the roller skating at the Wellington Pier and sometimes went to the dance at the Britannia Pier.

At the beginning of December, the *Fair Morn* headed home to Hopeman. After my first nine-week trip to Yarmouth, I remember my share of the pay-out was £26. At Hopeman harbour the nets were taken ashore and fresh nets were taken on board, ready for the West Coast.

In the second week of December, after a week at home, we sailed over to the west coast of Scotland, where we fished for herring out of Ullapool, Gairloch, Oban and Mallaig. Much of this time we were fishing in the Minches and around the Outer Hebrides. No-one seemed to bother much about Christmas in those days – in fact I can remember that on one Christmas Day at Castlebay (Barra), we landed one of our best ever catches. However, we always came home for Hogmanay, the New Year celebrations.

The weekends on the West Coast were nothing like as lively as those in Yarmouth. Stornoway and Oban had cinemas but they were closed on Sundays. Mallaig sometimes had a mobile cinema that came round in a van and showed films in a local hall. Other places had no entertainment at all except for bars.

Sundays were always quiet. Some of the men might go to the local kirk and some of us went for long walks but if it was raining we stayed on board and slept. Every three to four weeks we left the boat tied up in somewhere like Oban and got the bus or train back to Hopeman for a weekend.

On the west of Scotland the herring were not only found in the open sea but also in the long sea lochs and the skipper sometimes tried a different fishing routine. What we would do was anchor a string of drift nets in one of the sea lochs in the early evening and then take another thirty nets out to the open sea and shoot them in the normal way. After we had hauled these nets, we would return at about 4 a.m. to the loch to empty the anchored nets and then lay them out again. Then we headed for port. This was quite an exhausting way of working as we had no time to get any sleep at all. The following night we would just do the nets in the loch and the skipper would let us catch up on some sleep.

After the spring fishing we returned to Hopeman for a week before we went to Fraserburgh for the summer herring. Between 35 and 60 miles north east of Fraserburgh are a series of banks on the sea bed similar to those off Yarmouth. We based ourselves at Fraserburgh rather than Peterhead or Aberdeen because it was nearer to Hopeman and so we were able to go home most weekends.

From late August to September we fished off north-east England, mostly around Whitby and Scarborough, but we also made landings in North Shields and Hartlepool. We had to use nets with a slightly different mesh size there because the herrings there were a bit smaller than those off East Anglia so we had to go back home to change our nets again before heading for Yarmouth.

One year the Scarborough season lasted longer than expected and the skipper decided to give Yarmouth a miss and instead we headed to southern Ireland, where we fished out of Kinsale and Dunmore East. A couple of times we landed our catch at Milford Haven because you could get a better price there than in Ireland. From Ireland we sailed to Troon in Ayrshire to tie up the boat and go home for Hogmanay. Afterwards we went to Buncrana, County Donegal, for a few weeks before heading up to the Minches again.

I got worried one December night when we were fishing off Barra. We had got in among a good shoal of herring but by the time we had finished hauling, the weather had really deteriorated and as we headed for Oban we got caught in a gale and blizzard. It turned out to be the worst storm of that winter. After four hours in sleet and snow, running downwind across the Minch towards the mainland, there was no sign of the storm abating. The skipper decided that in the poor visibility and with no radar, it would be safer to turn back into the gale towards the Hebrides rather than risk getting blown on to rocks on the mainland. As he turned her about, the *Fair Morn* pitched so violently that we were all thrown out of our bunks and the cabin table overturned. It took us another ten hours battling in the teeth of the gale to reach safety at Castlebay.

The Post War Herring Trade

In the immediate post-war years, the herring trade was operating much as it had in the previous fifty years, with 'freshers', 'smokers', Scotch curers, canners and 'Klondykers' busy in the market. A reasonable proportion of the catch was exported, with Germany (both East and West) and the Soviet Union once again buying British salted herrings. In 1951 Mac Fisheries wound up its subsidiary Andrew Bremner Ltd, one of the oldest herring curing companies, and transferred the curing work to Bloomfield's.

Modern freezing techniques were now increasingly being used to even out peaks and troughs in the supply of herrings during the course of the year and to ensure continuity of supply for the herring processors. However, as frozen herrings became more widely available in northern Europe, the demand for salted herrings began to fall. UK exports of salted herring fell from 642,000 hundredweight in 1946 to 209,000 in 1960. The falling demand put the British curers under financial pressure and by 1960 the Board was helping to finance curing at the Scottish summer season and even underwrote the weekly pay of the gutters and coopers.

The creation in 1957 of the European Common Market, which at that time excluded the United Kingdom, did nothing to help British herring exporters. Because the Market abolished import duties between member states, it meant that there was a price advantage for Germany to buy Dutch herrings rather than British. To help keep costs down, the Board was keen to promote the use of recently-developed gutting machines and bought several at a cost of £850 each. The use of these machines led to fewer herring girls being needed. In 1959 the Board sanctioned the employment of just eleven gutting crews for the Yarmouth season, and the following year only three. Long gone were the days when there had been work for 5,000 girls at Yarmouth and Lowestoft.

The traditional export trade in hard-smoked herring recovered after the war when Greece and Italy started buying again. Indeed, this particular branch of the trade was deemed important enough in the 1950s for British Railways to introduce, for a trial period, a direct freight service carrying smoked herring by rail from Yarmouth right through to Italy. This was in addition to the daily trains carrying fresh herring from the Yarmouth and Lowestoft to the major cities (see Jack Stowers' story).

Despite the Board's best efforts, demand from the home market was now falling as housewives preferred to buy new products such as fish fingers, which, unlike herrings, required no preparation before cooking. Ironically, fish fingers were first produced in the United Kingdom by Birds Eye (in 1955) at Yarmouth, the home of the herring industry.

Two new branches of the herring trade emerged in the 1950s – pet food and marinating. In 1960, at Stornoway alone some 6,500 crans of herring were sold to pet food manufacturers. The marinating of herring was based on Scandinavian herring recipes. The fish were packed in a mixture of vinegar, brown sugar and spices and were sold variously as 'pickled', 'soused' or 'roll-mop' herrings.

In the search for new products and new markets, British Kipper Exporters Ltd (a marketing co-operative set up by the Herring Industry Board) went all-out to promote sales in the United States. In 1950–51 they produced 320 tons of kippers which they gave the brand name 'Queen of Scots'. They shipped them across the Atlantic but they were not a success and a substantial number remained unsold. Undaunted, in 1953 they introduced the United States consumers to the 'Edinburger', the world's first fresh-frozen boneless kipper. These were cooked, formed into slabs and then frozen. The American consumer could buy a box of twenty-five Edinburgers for just 59 cents. From the fact that, after 1953, there was no further mention of them, it can be assumed America was not too impressed.

In 1960 the Board tried a new way to sell kippers in the home market when it conducted trials with a coin-operated vending machine which dispensed packs of quick-frozen kipper fillets. This scheme rather smacks of desperation but although some kippering companies showed interest, there was insufficient enthusiasm from the public to warrant promoting the idea to retailers.

North Sea Herring Stocks and Catches

Some 90 per cent of the 1950s East Anglian herring landings were still being caught using drift nets which only caught adult fish. Elsewhere in the North Sea, vessels from other countries (Denmark in particular) were trawling immature herring and other pelagic species to supply their fish reduction factories. Most vessels were now using echo-sounders to locate the shoals of herring. French and Belgian trawlers compounded the pressure on herring stocks by trawling the sea bed off Calais and Dunkirk where the herrings were spawning on the Sandettie shallows, scooping up the adult fish and damaging newly-laid eggs.

There appears to have been little willingness to co-operate in conservation of the herring stock. The destruction of herring spawn and immature young fish meant that each year there were fewer herrings of breeding age to maintain the stock. It was a vicious cycle, and with no intervention the inevitable began to happen.

In both England and Scotland herring catches fell after the war.

	1938	1950	1960	1966
England & Wales	2,574,455	1,528,841	332,164	256,000
Scotland	2,801,000	1,997,273	1,762,000	1,997,485

Table 18 Falling herring catches, 1950s.

The British herring industry had in the past experienced occasional years when shoals were scarce and catches were low, but what it was witnessing now was a relentless fall in herring numbers, particularly in the southern North Sea. In its annual report of 1960, the HIB resignedly stated: 'There was nothing in the experience of 1960 to suggest that nature is succeeding in counteracting the effects of man's onslaught on the North Sea herring stocks at all stages of their growth.' In Scotland, as post-war herring landings started to fall at the traditional North Sea herring ports (Peterhead, Fraserburgh and Wick), landings at the west coast ports of Oban, Mallaig and Ullapool were increasing (see Appendix 7). In the late 1960s Mallaig would for a while take over Yarmouth's former title as the 'Herring Capital of Europe'.

From 1950, the number of fishing boats (both English and Scottish) joining the East Anglian herring fishery steadily fell. At Yarmouth, the post-war numbers peaked in 1949 at 302 vessels but it was becoming harder to land good catches off East Anglia. Despite this, in Yarmouth and Lowestoft there was still prosperity and confidence in the industry. The Prunier Trophy continued to produce some staggeringly high landings, although some of these were achieved by using an incredible four miles of nets. In the early 1950s a similar herring competition, for the Boothby Trophy, was inaugurated in north-east Scotland and was held each summer until 1966. (Robert Boothby continued to represent the local constituency until 1958, when he was made a life peer.) In 1958 Madame Prunier ended her direct involvement with the competition which carried her name and handed the responsibility to the HIB.

Bloomfield's were among several Yarmouth and Lowestoft companies who throughout the 1950s continued to operate some quite elderly steam drifters, although they were gradually replacing them with diesel powered drifter/trawlers. Owners had really had their money's worth out of the old steamers, some of which were now thirty years old. Once they reached the point that major expenditure was required on them, they were sold off, usually for scrap. In Scotland, where the diesel powered boat had long reigned supreme, the disappearance of steam drifters was more rapid and by 1955 there was just one remaining on the Scottish register (a Peterhead boat) and that had gone by the following year. As late as 1958, Bloomfield's was still buying coal in Durham and shipping it to Wick to supply English steam drifters.

Year	England & Wales	Scotland
1938	485	402
1950	147	190
1955	60	1
1960	5	0

Table 19 Steam drifter numbers, 1938–60.

In 1960 there were only forty-two drifters remaining on the registers at Yarmouth and Lowestoft, of which just five were steam-powered. 1961 saw the withdrawal of the last steel-built drifter, *Prime*, LT77). That year, wood-built *Wydale*, YH109, briefly became the last steam drifter still operating before she too was withdrawn. The future for herring fishing was already beginning to look bleak even as in 1960 the last drifter/trawler to be built in the United Kingdom was launched for Small & Co. at Lowestoft. In late 1962 Mac Fisheries decided that there was little economic future in herring fishing and sold the five remaining Bloomfield's diesel drifter/trawlers to Small and Co.

Postscript

The Great Yarmouth Boat Owners' Association held a meeting in February 1964 at which they acknowledged that there were no longer any locally-owned drifters, and so this became their final meeting. The town's Fishermen's Widows and Orphans

Fund was wound up in 1967, at which time fifteen widows and four dependent children were receiving an average of fifteen shillings a week. At Lowestoft in 1967 the once powerful English Herring Catchers' Association held its final annual general meeting with just three members present.

As the herring disappeared from the North Sea, ports like Lowestoft, Fraserburgh, Aberdeen, Peterhead and North Shields continued to prosper through landings of whitefish. In the late 1960s both Lowestoft and Yarmouth became shore bases for the new oil and gas exploration industry in the southern North Sea. Indeed, some of the diesel drifter/trawlers formerly owned by Bloomfield's and Smalls were now given a new lease of life as rig safety stand-by vessels. Mac Fisheries closed down what was left of Bloomfield's in 1977 and two years later Unilever disposed of the Mac Fisheries retail chain.

In Scotland today there are some twenty pelagic trawlers which catch herring to the north and west of Scotland. Their catches are landed at Lerwick, Fraserburgh and Peterhead, all once famous as herring stations. All of the herrings landed there these days are destined for human consumption (no more reduction to oil or animal feed). The amount of herring caught is governed by European quotas and in 2010 the total (official) landings in Scotland were 40,000 tonnes with a value of nearly £12 million. Impressive as these figures seem, they are only a third of the annual landings at those three ports in the late 1920s.

Former fishing harbours which once heaved with herring drifters now host a few small inshore fishing boats and dozens of recreational yachts and speedboats. The number of people who can recall work in the herring industry or even seeing it at

HS Fishing of Yarmouth keep up the traditional export trade in hard smoked herring to Mediterranean countries. Note the elegant cowls at the top of the smokehouse.

work is gradually decreasing. Sadly, many buildings which were once an important part of the herring industry have disappeared but a few survive, often converted to other uses.

Around the British coast smokehouses still produce kippers and bloaters for the home market, although in order to maintain production they often have to buy Norwegian herrings. In Yarmouth, HS Fishing 2000 Ltd (occupying the premises once owned by Henry Sutton) continues the centuries-old tradition of producing hard-smoked herring for the export market. They sell Golden and Silver Herrings (smoked for about a week) to Italy, Greece, Cyprus and Kuwait. They also produce a hard cured herring, which is smoked for three months (yes!) and is popular in Africa. The current company has four employees – at the height of the industry, Henry Sutton employed 170.

The most notable survivor from the great herring era is the steam drifter *Lydia Eva*. Restored to working order with the help of the Heritage Lottery Fund, she now spends the summer months moored opposite Yarmouth Town Hall, at the spot where, many years ago, hundreds of Scottish drifters would tie up for the weekend. Here, a group of knowledgeable volunteers welcome visitors and help them to learn the story of the herring fishing. With her boiler fired up, she makes a fine sight as she heads out to sea under her own steam, the last surviving example of a vessel that a century ago helped transform the British herring industry.

Timelines Chapter 9

1945 The Danes resume industrial herring trawling in the North Sea
1945 The branding of barrels of herring is phased out
1946 The first herring reduction factories in the UK are opened
1946 The Prunier Trophy returns after the war
1946 300 boats work the East Anglian season
1949 Wee Green dies aged 72
1949 Scotch girls go on strike in Yarmouth for higher ration allowance
1952 A herring reduction plant is opened at Yarmouth
1954 'Painted Ladies' start to disappear from the kipper trade
1955 Birds Eye begin producing fish fingers in Yarmouth
1960 The last drifter/trawler to be built is launched at Lowestoft for Small & Co.
1960 Five steam and thirty-seven motor drifter/trawlers still on the register at
     Yarmouth and Lowestoft
1961 The last steel steam drifter (*Prime*, LT77) is withdrawn from service
1961 The last wood-built steam drifter (*Wydale*, YH105) is sold for scrap
1963 Bloomfield's (Unilever) sell their last six drifter/trawlers to Small & Co. of
     Lowestoft
1965 Great Yarmouth's herring reduction plant is demolished

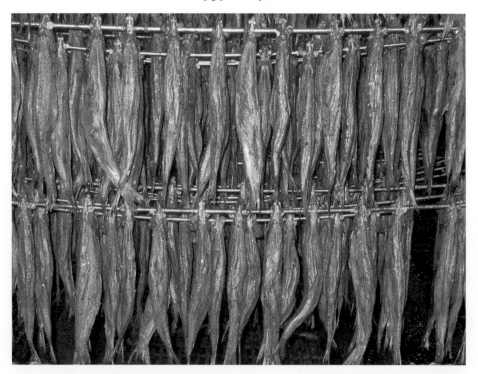

Racks of herrings begin another day in the smokehouse. These herring are destined for Cyprus.

The steam drifter *Lydia Eva* was built in 1930 and she fished until 1938. She then filled a variety of support roles for the RAF and the navy and survived in running order until preservation in 1969.

1966 The Prunier Trophy is wound up

1967 2,600 crans of herring landed at Great Yarmouth (compared to 824,000 in 1913)

1968 The final year that a herring drifter (*Wisemans*, LT382) fishes out of Lowestoft

1970 Seventy-one crans of herring landed at Great Yarmouth (by ring-netters)

1977 Unilever wind up Bloomfield's Limited

1979 Unilever close down Mac Fisheries retail chain

Widows and Orphans' Fund

Name of Claimant	Relationship to Deceased	Deceased's Name	Name of Vessel	Death Grant	Weekly payment
Mr W. Steel	Father	Arthur Steel	*Blackthorn*	£10	–
Caroline Brown	Mother	Thomas Brown	*Ludham Castle*	£10	–
Eliza Davis	Wife	William Davis	*Twenty-Eight*	–	7s 0d
Mrs C. Woolston	Mother	Alexander Woolston	*Irene*	£10	–
Eliza Dyble	Wife	Robert Dyble	*Montrose*	–	13s 0d
Silas George	Father	Silas George Jun.	*Montrose*	£10	–
Joseph Julien	Father	Sydney Julien	*Montrose*	£10	–
Charlotte Addy	Sister	Not stated	*Kestrel*	£10	–
Florence Balls	Wife	Not stated	*Kestrel*	–	13s 0d
Harriet Guyton	Wife	Benjamin Guyton	*Kestrel*	–	7s 0d
Phoebe Bartram	Wife	Thomas Bartram	*Kestrel*	–	7s 0d
Susannah Harlow	Wife	Francis Harlow	*Kestrel*	–	7s 0d
Eliza Harris	Mother	L. T. Harris	*Kestrel*	£10	–
Eliza Thompson	Wife	Thomas Thompson	*Kestrel*	–	7s 0d
The Rector of Carleton Rode (on behalf of deceased's mother)		Arthur Edwards	*Montrose*	£10	–
Wm Knights	Father	Frederick	*Montrose*	£10	–

Exports of
British Cured Herring

Destination	1913 (Cwts)	1928 (Cwts)	1938 (Cwts)
Argentina	–	960	1,716
Australia	–	7,096	4,513
Belgium	222,505	173,541	16,009
British West Africa	–	733	1,201
Canada	–	18,749	4,615
Cyprus	–	3,246	3,855
Egypt	–	28,766	47,497
Estonia	–	124,002	90,971
Finland	–	143,291	25,278
France	–	12,868	10,014
Germany	3,996,892	1,944,001	971,656
Greece	42,169	47,269	54,801
India	–	952	1,596
Italy	91,590	135,955	70,440
Latvia	–	799,922	174,734
Lithuania	–	43,800	167,391
Malta	–	4,966	4,323
Netherlands	167,641	107,833	12,792
New Zealand	–	1,142	619
Norway	38,593	4,100	–
Palestine	–	4,527	6,772
Poland (inc. Danzig)	–	1,523,701	864,960
Romania	–	–	15,488
Russia	3,566,155	102,285	76,026
South Africa	–	17,305	14,975
Straits Settlements	–	1,072	713
Sweden	–	7,560	10,961
United States	398,810	168,987	56,521
Other Countries	88,571	21,748	25,625
TOTALS	8,612,926	5,450,377	2,736,062

Source: Annual Report on Sea Fisheries

The Russian Northern Industries Investors, 1913

Name	Investment (Roubles)
Vladimir de Sivers	50,000
Captain Carle J. Spahde	5,000
Grooten	25,000
Baroness Palmia	25,000
Mera	25,000
James Sivers	5,000
Theodor Sivers	5,000
Olga Sivers	5,000
Loretz Eblin	15,000
Admiral Aboze	5,000
Vice Admiral Rowlands	10,000
Captain Islomic	5,000
General Duckie	10,000
Brandt	15,000
Schron	5,000
Sangovitch	5,000
Lemonius	5,000
Bloomfield's Limited	110,000
Total	330,000

Fishing Vessel Insurance Claims

Date	Vessel	Location	Nature of mishap
21/06/1923	*Angelina*	Lerwick	Damaged in collision
31/07/1923	*Speranza*	Lerwick	Broken tail shaft
31/05/1924	*Lord Fisher*	Newlyn	Struck rock and broke all blades on propeller
22/08/1924	*Majesty*	Hartlepool	Sank on rocks
22/10/1924	*Nine Sisters*	Lowestoft	Broken mizzen mast
27/03/1925	*B J B*	Scillies	Damaged in collision with unknown schooner
27/03/1925	*Adele*	Padstow	Stern damaged in harbour
27/05/1925	*Ben and Lucy*	Newlyn	Struck Newlyn Quay
11/06/1925	*Boy Scout*	Hartlepool	Broke keel on Parten Rocks
11/09/1926	*Golden Spur*	Scarborough	Caught fire and sank
31/03/1927	*Ben and Lucy*	Newlyn	Damaged in heavy gale
19/05/1927	*John Alfred*	Shields	Struck and sunk by Latvian collier
11/08/1927	*Ben and Lucy*	Shields	Broken shaft
26/01/1928	*Harnser*	Buncrana	Grounded in harbour
18/07/1928	*Three Kings*	Stronsay	Broken propeller
17/09/1928	*United Boys*	Caledonian Canal	Broken propeller

Checklist for a Herring Girl's Kist

A thin mattress
Sheets, pillow and pillow case
Blankets amd towels
Six Shifts (vests)
Six drawers (knitted knickers)
Work clothes – coat, skirt, gansey, aprons, woollen stockings, shawls, gumboots
Gutting knives, sharpening stone
Sunday best – blouse, shoes, skirt, cardigan, coat and hat
Knitting needles, wool and knitting pouch
Toiletries – soap, talc, hairbrush, hairgrips
Cutlery, mug, plate
A bible

Herring Catches (England and Wales)

Port	1913 (Cwts)	1928 (Cwts)	1938 (Cwts)
Berwick	58,921	11,442	235
Blyth	155,643	–	–
North Shields	500,718	223,310	159,654
Hartlepool	154,542	94,221	26,591
Whitby	–	534	72
Scarborough	128,124	534	6,137
Hull	203,256	35,747	121,215
Grimsby	406,705	125,099	39,886
Cromer	1,565	981	312
Winterton	–	31	1,535
Yarmouth	3,122,282	1,964,392	1,137,784
Lowestoft	2,151,463	1,296,484	859,755
Folkestone	6,864	1,268	–
Hastings	15,031	5,862	3,677
Newhaven	26,784	3,845	236
Plymouth	23,037	112,215	566
Newlyn	–	7,851	3,760
St Ives	5,328	19,718	1,859
Milford Haven	23,722	93,125	89,204
Liverpool	14,192	–	–
Fleetwood	66,255	46,664	76,080
Peel, I.O.M	110,253	32,388	27,784
Port St Mary I.O.M	48,344	12,814	–
Other Stations	63,856	67,485	39,896
TOTALS	7,286,885	4,156,010	2,596,238

Source: Annual Report on Sea Fisheries

Herring Catches (Scotland)

Port	1929	1949	1955	1960
Eyemouth	68,622	2,242	984	42
Newhaven	46,346	51,295	5,844	13,466
St Monance	45,080	80	256	–
Anstruther and Cellardyke	61,481	300	–	–
Aberdeen	19,487	297,945	259,712	119,022
Peterhead	543,599	164,564	315,561	77,289
Fraserburgh	607,612	262,731	606,530	245,741
Macduff	26,485	1,476	52	–
Buckie	25,412	7,475	1,133	7,540
Lossiemouth	3,496	329	–	–
Inverness	5,810	13,992	161,634	17,802
Wick	375,994	39,149	25,187	62
Stronsay/Orkney	211,343	–	–	–
Lerwick	1,034,203	235,045	148,573	104,313
Stornoway	346,882	207,572	109,690	121,766
Castlebay	120,961	332	56	52
Ullapool	–	356,877	143,243	262,570
Gairloch	–	–	99,746	67,977
Mallaig	156,224	156,500	174,233	131,040
Oban	92,544	92,544	140,707	193,559
Campbeltown	1,040	57,485	11,647	282
Tarbert	3,003	280,905	28,343	148,188
Rothesay	87,802	623	–	–
Wemyss Bay	–	111,352	–	–
Ayr	765	183,720	27,573	152,627
Portpatrick	218	47,813	79,752	–
Other harbours and stations	169,395	17,781	66,377	98,720
TOTAL	4,053,804	2,590,127	2,406,833	1,762,058

Source: Annual Report on Sea Fisheries

Further Information

Hopefully, this book has whetted your appetite to find out more about the British herring industry. Here is a selection of books, museums and websites which will be of interest.

Recommended Books

Anson, Peter F., *Fishing Boats and Fisher Folk on the East Coast of Scotland*, J & M Dent & Sons Ltd, 1930

Butcher, David, *The Driftermen*, Tops'l Books, 1979

Butcher, David, *Following the Fishing*, Tops'l Books, 1987

Butcher, David, *Living from the Sea*, Tops'l Books, 1982

Coull, James R., *The Sea Fisheries of Scotland*, John Donald Publishers Ltd, 1996

D'Enno, Douglas, *Fishermen Against the Kaiser*, Vol. 1, Pen and Sword Maritime, 2010

Frank, Peter, *Yorkshire Fishing Folk*, Phillimore, 2002

Gray, Malcolm, *The Fishing Industries of Scotland 1790–1914*, University of Aberdeen, 1978

Hawkins, L. W., *The Ocean Fleet of Yarmouth*, 1983

Hodgson, W. C., *The Herring and its Fishery*, Routledge and Keegan Paul, 1957

Jenkins, James T., *The Herring and the Herring Fisheries*, PS King & Sons, 1927

Martin, Angus, *The Ring-Net Fishermen*, John Donald, 1980

Nazé, Yannick, *La Pêche au Hareng*, Editions des Falaises, 2010

Smylie, Mike, *Herring: A History of the Silver Darlings*, Tempus, 2004.

Starkey, D. J., Reid, C. and Ashcroft, N. (Eds), *England's Sea Fisheries*, Chatham Publishing, 2000

Tarvit, Jim, *Steam Drifters: A Brief History*, St Ayles Press, 2004

Telford, Susan, *In a World a Wir Ane*, The Shetland Times Ltd, 1998

Tooke, Colin, *The Great Yarmouth Herring Industry*, Tempus, 2006

White, Malcolm, *Herrings, Drifters and the Prunier Trophy*, 2006

Wilcox, Martin, *Fishing and Fishermen: A Guide for Family Historians*, Pen & Sword, 2009

Places to Visit

The following places are recommended by the author. If you are planning to visit any of them, please check their websites as some have limited opening times.

Buckie and District Fishing Heritage Centre, Moray
A small, volunteer-run museum containing lots of fascinating artefacts and documents relating to the Moray Firth herring fishing.
www.buckieheritage.org

Great Yarmouth Potteries, Norfolk
This working pottery, established in a former smokehouse built into the medieval town wall, houses a herring smoking museum packed with images and artefacts from the heyday of the herring industry.
www.greatyarmouthpotteries.co.uk

Lossiemouth Fisheries and Community Museum, Moray
A volunteer-staffed museum situated on the quayside at Lossiemouth harbour and full of fishing artefacts. It also has a section dedicated to local hero Ramsay MacDonald.
www.lossiemuseum.co.uk

Lowestoft Maritime Museum, Suffolk
A volunteer-run museum that is packed full of artefacts and pictures from both the drifting and trawling industries; it houses the Prunier Trophy and the official Cran measure.
www.lowestoftmaritimemuseum.org.uk

Lydia Eva, Great Yarmouth, Norfolk
The last remaining steam drifter, built in 1930, is a must if you are visiting Great Yarmouth. She is normally moored there from May to October and is open to visitors.
www.lydiaeva.org.uk/lydia_eva

Norfolk Record Office, Norwich
This archive holds the Great Yarmouth port records and also the business records of Bloomfield's Limited – ledgers and minutes of board meetings.
www.archives.norfolk.gov.uk

Scottish Fisheries Museum, Anstruther, East Fife
A museum dedicated to the Scottish fishing industry, situated on the harbour at Anstruther in Fife. It houses a fine collection of boats and fishing equipment.
www.scotfishmuseum.org

Southwold Museum, Suffolk
A small museum with a section dedicated to fishing. It has a very informative website with information on the local herring industry.
www.southwoldmuseum.org

Time and Tide Museum, Great Yarmouth, Norfolk
This modern museum has been established in a former herring smokehouse. It tells the story of Yarmouth and the herring industry.
www.museums.norfolk.gov.uk/Visit_Us/Time_and_Tide

Wick Heritage Museum, Caithness
A museum that covers the history of Wick in which the herring industry features prominently. An excellent collection of fishing and coopering artefacts.
www.wickheritage.org/

Useful Websites

There is a wealth of information and images available on the internet which can be found by using a search engine. However, I recommend the following as good starting places:

British Pathé
Use the Search facility on this website to view a wonderful selection of newsreel films from 1920 to 1960 of British herring fishing and gutting.
www.britishpathe.com

East Anglian Film Archive
This on-line archive of films made by amateurs and professionals includes a good selection of fishing and herring-related films and TV programmes from Lowestoft and Yarmouth.
www.eafa.org.uk/

Johnston Collection
Visit this website to view the remarkable collection of old photographs from Wick and Caithness, including many images of the herring industry.
www.johnstoncollection.net

Maritime History Archive, Memorial University of Newfoundland
This Canadian archive hold crew lists for a great many British fishing vessels and for a fee will issue copies, but you do need to know the official number of the vessel as well as the name.
www.mun.ca/mha/index

Orkney Image Library
Very good selection of photographs of the Stronsay herring industry
www.orkneycommunities.co.uk/imagelibrary

Penlee House Gallery and Museum
This Penzance gallery is itself worth a visit if you are in Cornwall; if not, their website has a good selection of old photographs of fishing at Newlyn.
www.penleehouse.org.uk/collections/photography

Picture Norfolk
This website is maintained by Norfolk Library Service and has many herring-related images, mainly from Great Yarmouth.
www.picture.norfolk.gov.uk

Fisheries Statistics Archives
Statistical returns for the English fishing industry (and some for the British industry) 1866 to date.
www.marinemanagement.org.uk/fisheries/statistics/annual_archive

Statistical returns for the Scottish fishing industry 1927 to date.
www.scotland.gov.uk/Topics/Statistics/Browse/Agriculture-Fisheries/PubFisheries

Glossary

Allotment: Weekly advance paid by English fishing companies to crewmen's families while the men were away on a voyage.

Arles: Engagement fee paid by a herring curer to herring girls to seal the contract for a season's work.

Barking: Scots term for 'tanning'.

Barrel money: The main part of the herring girls' pay, a piece-work payment to a crew for each barrel of herrings they completed.

Beatster: Woman employed to repair drift nets (mainly East Anglia).

Bloater: An ungutted herring dipped in brine for a couple of minutes and then smoked for about seven hours.

Bowl: Another name for a 'buff'.

Buff: A large float, usually painted white, which supported the driftnets in the sea and marked out a train of nets.

Buoy: Another name for a 'buff'.

Capstan: Winch used for hauling in ropes and nets.

Cloots: Strips of cloth which the herring gutters tied on their fingers in order to protect them from cuts.

Coal hulk: Floating coal store, usually a decommissioned cargo boat.

Coalie: Labourer employed to carry sacks of coal from the coal lorry on to a steam drifter.

Cran: Volumetric measure of herrings, 37.5 gallons, equating to 28 stones in weight.

Crew: Working unit of three herring girls, comprising two gutters and a packer.

Crown brand: Set of quality standards for Scottish salted herring, evidenced by the official marking ('branding') with a crown of each full barrel.

Cured herring: Another name for 'salted herring'; sometimes smoked herring are also referred to as cured herring.

Curer: Entrepreneur, usually Scottish, who employed gutters and coopers to produce barrels of cured herrings.

Curing yard: Area of land close to a harbour which a curer would hire as his business premises.

Cutch: Extract from South East Asian acacia trees which was dissolved in tanks of boiling water and used for tanning nets.

Demersal: The class of sea fish that live permanently on the sea-bed, for example, cod, haddock, plaice and sole.

Double swim: When a net is hauled and is found to have herrings on both sides, indicating that herring have swum into it from opposite directions.

Drifter: Fishing vessel, powered by sail, steam or motor, which used drift nets.

Drift net: Type of gill net, a large area of net suspended, curtain-like, close to the surface of the sea into which pelagic fish swim and become stuck.

Driver: Another term for an 'engineer'.

Engineer: The man in charge of the engine and all things mechanical on a steam drifter and responsible for providing power to the propellers and the capstan.

Farlan (also 'farlin' or 'farlane'): A trough-like workbench at which the herring gutters stood and worked.

Fireman: Crewman whose main responsibility was to keep the furnace stoked with coal, to maintain steam pressure in the boiler.

Fo'c's'le: The forward part of a vessel.

Fresher : Merchant who traded in fresh herrings

Gypping: Gutting

Gutting: Removing the gills and alimentary tract of the herring with a small, sharp knife (but leaving the roes in situ).

Hawseman: Drifter crewman ranked immediately below the mate (East Anglia).

Herring station: Harbour used by herring boats where the herring were processed for export.

Home Fishing: Name given by Yarmouth and Lowestoft fishermen to the East Anglian autumn herring season.

Kipper: A gutted fresh herring, split open down its back, lightly salted and smoked for six to eight hours.

Kipperer: A 'smoker' who specialised in kippers.

Kist: Wooden trunk in which a herring girl would pack all she would need for ten weeks working away from home.

Klondyke: To pack fresh, whole herrings with ice and salt into wooden cases for export, usually to Germany.

Klondyker: Merchant (or a ship) involved in the 'Klondyke' trade.

Last: Ancient measure of herring landings in which the fish were manually counted; a last equals 1,320 herrings.

Layer off: Another name for a 'coalie'.

Leader: Scottish name for the 'warp'.

Look on: Exploratory haul of the first couple of nets to see if there were any herring in them.

Messenger: Another name for the 'warp'.

Mizzen: Mast in the stern of a drifter, and also the small sail attached to it.

Overdays: Herrings which were not landed on the day they were caught but were salted on board while the boat stayed out another night.

Painted lady: Disparaging term used for kippers which were painted with red dye before being smoked.

Pallet: Another name for a 'buff'.

Pelagic: Class of sea fish that rise to the surface to feed on plankton; includes herring, mackerel, sprats and sardines.

Pickled herring: Another name for 'salted herring' (nothing to do with vinegar!).

Pickling liquor: The distinctive brine produced in barrels of salted herring by the salt absorbing blood and moisture from the herrings.

Pickling plot: Another name for a 'curing yard'.

Pierhead painter: Artist who in the early 1900s made a living by producing paintings of vessels, usually selling them to fishermen.

Ransacker: Man ashore who checked driftnets to assess what repairs were needed.

Ring-net: Very large net used to surround herring and draw them to the surface.

Rouse: Mix salt into fresh herring with a shovel, before gutting began.

Salted herring: Fresh herring, gutted and packed tightly with salt into wooden barrels. A popular food in Russia and Germany, they would keep for six months.

Scud: To shake the nets during hauling to dislodge the herrings from them.

Scotch cure: The strictly controlled process followed by the Scots curers in producing salted herring for export.

Scotch Voyage: The name given by East Anglian driftermen to the herring season in northern Scotland which began each June.

Scutcher: Large scoop used in the hold of a drifter to shovel herrings into baskets for unloading.

Scuttle: Small opening in the deck of a drifter through which herring slid down into the hold (like a small manhole).

Seine-netting: Type of trawling which the Scots copied from the Scandinavians in the 1920s.

Shimmer: The first indication from a 'look on' of the possible size of a catch of herring (e.g. "a forty cran shimmer").

Smoker: A merchant involved in producing smoked herring

Spent: Adult herring in its post spawning condition.

Stocker: Demersal fish caught by driftermen using hand lines during any spare time on board – a means of earning some beer money before pay day.

Stoker: More usual term for 'fireman'.

Swill: Distinctive double basket used for transporting herring from quayside to nearby yards and smokehouses (unique to Yarmouth).

Tanning: Dipping drift nets into a solution of 'cutch' in order to protect them from the damage caused by prolonged exposure to sea water.

Trawling: Type of fishing in which a large net is dragged along the sea bed to trap fish inside (rather like a string shopping bag).

VAD: Title by which Red Cross nurses were known, an acronym for Voluntary Aid Detachment.

Warp: Main rope, two to three inches thick and a mile and a half long, to which drift nets were attached.

Westward: Name given by East Anglian fishermen to the winter or early spring voyage to Devon and Cornwall, fishing for herring and mackerel.

Whaleman: Grade of drifterman between the hawseman and the younker (East Anglia).

White herring: Another term for 'cured herring'.

Younker: Yarmouth name for the crewman who job was to detach the nets from the warp during hauling.

Acknowledgements

I would like to thank Vernon Green, Jack Stowers, Gerald Tungate, Sandy McPherson, Chrissie Bruce and Isobella Christie for patiently answering my stream of questions. I am also indebted to Peter Allard, Peter Jones, Andrew Fakes, James Fisk and Claire Everitt for their help.

Picture Credits

Thank-you to the following individuals and organisations for granting permission to use their images: Archant, p.48 right; Barbara Pilch, p.109 left; Bloomfield's Archive (James Fisk), p.35, p.45 top, p.80, p.83; Campbell McCutcheon, p.9, p.10, p.28, p.31 top, p.119; *Fishing News*, p.19, p.48 left, p.77, p.123; Geoff Moore, p.131; George Plunkett Collection, p.86; Graham Kenworthy, p.100 bottom; Jack Holmes Collection, p.113; Jenny Unsworth, p.85; Leece Museum, Peel, p.117; Norfolk County Council (Picture Norfolk), p.39 bottom, p.73, p.78, p.104, p.106, p.133, (Time and Tide Museum), p.54; Norfolk Record Office (ref BR 51/106), p.97 top and bottom; Peter Allard, p.55. p.59; Peter Jones Collection, p.22, p.110 top and bottom, p.112; Sandy McPherson, p.134, p.135; The Scottish Fishermen's Organisation, p.100; The Scottish Fisheries Museum, p.121; Sarah Smith, p.116; *The Northern Scot*, p.87; Vernon Green, p.53.

ALSO AVAILABLE FROM AMBERLEY PUBLISHING

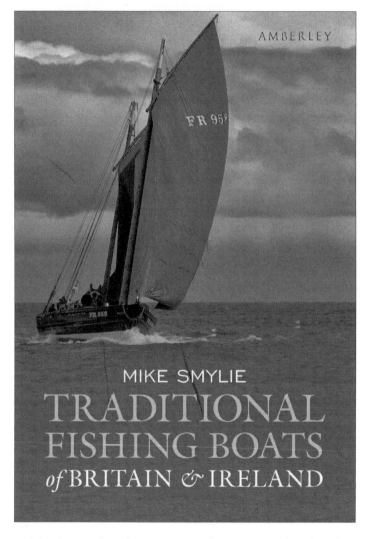

Traditional Fishing Boats of Britain and Ireland
Mike Smylie

Fishing once employed hundreds of thousands, with small harbours
all around the coast full of fishing boats. Mike Smylie takes us
on a tour round Britain and Ireland, describing the boats once in
common usage.

978 1 4456 0252 3

320 pages, 200 illustrations

Available from all good bookshops or order direct
from our website www.amberleybooks.com